AF618537

SpringerWienNewYork

CISM COURSES AND LECTURES

Series Editors:

The Rectors
Friedrich Pfeiffer - Munich
Franz G. Rammerstorfer - Wien
Jean Salençon - Palaiseau

The Secretary General
Bernhard Schrefler - Padua

Executive Editor
Paolo Serafini - Udine

The series presents lecture notes, monographs, edited works and proceedings in the field of Mechanics, Engineering, Computer Science and Applied Mathematics.
Purpose of the series is to make known in the international scientific and technical community results obtained in some of the activities organized by CISM, the International Centre for Mechanical Sciences.

INTERNATIONAL CENTRE FOR MECHANICAL SCIENCES

COURSES AND LECTURES - No. 533

MECHANICS OF CRUSTAL ROCKS

EDITED BY

YVES M. LEROY
LABORATOIRE DE GEOLOGIE, CNRS, ECOLE NORMALE SUPERIEURE,
PARIS, FRANCE

FLORIAN K. LEHNER
UNIVERSITY OF SALZBURG
AUSTRIA

SpringerWienNewYork

This volume contains 42 illustrations

This work is subject to copyright.
All rights are reserved,
whether the whole or part of the material is concerned
specifically those of translation, reprinting, re-use of illustrations,
broadcasting, reproduction by photocopying machine
or similar means, and storage in data banks.
© 2011 by CISM, Udine

SPIN 80084068

All contributions have been typeset by the authors.

ISBN 978-3-7091-0938-0 SpringerWienNewYork

PREFACE

The mechanical behaviour of the Earth's upper crust enters into a great variety of questions in different areas of the geological and geophysical sciences as well as in the more applied geotechnical disciplines. Different aspects of it are addressed by a continuously growing diversity of theoretical approaches ever since the days of Newton, Laplace, or Coulomb, some of which are now classical, while others are new and in development, with promising applications in sight. These studies of Earth deformation processes continue to contribute significantly to theoretical and applied mechanics and deserve the attention of students and researchers. A number of years ago, it was therefore thought timely by the editors of this volume to coordinate a lecture series on the subject Deformation of the Earth's Upper Crust - Theory, Experiment, and Modelling as an Advanced School at CISM Udine. A selection from the lectures held in June 2002 has now been brought up-to-date where necessary for publication in the present volume. While each of these chapters will hopefully make for a useful contribution in its own right, the present bundle can also illustrate, by way of examples, the variety of theoretical concepts and tools that are currently brought to bear on Earth deformation studies, ranging from reviews of poroelastic field theory in Chapters 1 & 2 to micromechanical and homogenization studies in Chapter 3, chemomechanics and interfacial stability theory of soluble solids under stress in Chapter 4, and finally, in Chapter 5, to an introduction to the Finite Element method as the most widely applied computational tool in geomechanics.

Chapter 1 will be of interest, in particular, for the formulation of experimental strategies for the measurement of poroelastic constants. An understanding of the various alternative formulations of the set of linearized constitutive relations is needed for this purpose and may be acquired by studying this lecture.

Chapter 2 deals with an important technique, due to Eshelby, for determining the stress and strain in regions of an infinite elastic body (with elastic constants that differ possibly from those of its surroundings), which undergo a change of size or shape due to loading in the farfield. Example applications of the technique include determining the effective stress in a narrow fault zone and the stress and strain in and around a fluid reservoir as generated by fluid mass injection or withdrawal. It also discusses the use of Eshelby's approach in seismic source theory.

Upper crustal rocks often contain cracks of diverse sizes, shapes and orientations. Chapter 3 deals with methods of predicting their influence on the effective elastic properties of a rock, these being of particular interest to seismologists. The theoretical methods discussed rely on the identification of microstructure-sentive parameters in terms of which the effective elastic constants of such rocks may accurately enough be expressed. The lecture offers a critical assessment of this problem. It also addresses the possibility of extracting from elastic wave velocities microstructural information on pore and crack content as well as on the presence and properties of pore fluids. The lecture thereby proposes to bridge a gap between geophysics and the general mechanics of materials.

The chemomechanics of pressure solution phenomena in solid/liquid systems under non-hydrostatic stress had been covered in a number of lectures by Profs. Spiers of Utrecht University and Lehner of the University of Salzburg at the CISM Advanced School in June 2002. Chapter 4 reviews the basic elements of a thermodynamic theory of pressure solution processes and covers recent developments by J. Raphanel on the evolution of fluid-infiltrated grain boundaries under stress. The lecture also offers an opportunity to study the linearization argument that forms a point of departure in most stability and bifurcation analyses.

Chapter 5 contains an exposition of the finite-element method for elastic and elasto-plastic materials under quasi-static conditions. Although this lecture had not been given at the 2002 CISM School it is added here to the theoretical topics of the preceding chapters in order to do justice to the practical need for approximate methods of solution by numerical means. This chapter grew from lectures taught by its author at the Ecole Polytechnique at Palaiseau, the California Institute of Technology at Pasadena, and the Ecole Normale Supérieure in Paris.

Y.M. Leroy
F.K. Lehner

CONTENTS

The Linear Theory of Anisotropic Poroelastic Solids

Florian K. Lehner

Department of Geography and Geology, University of Salzburg, Austria

Abstract This chapter offers a comprehensive derivation of the constitutive equations of linear poroelasticity. A main purpose of this survey is to assist with the formulation of experimental strategies for the measurement of poroelastic constants. The complete set of linearized constitutive relations is phrased alternatively in terms of undrained bulk parameters or drained skeleton parameters. Displayed in the form of a mnemonic diagram, this can provide a rapid overview of the theory. The principal relationships between alternative sets of material parameters are tabulated, among them the well-known Gassmann or Brown-Korringa fluid substitution relations that are rederived here without any pore-scale considerations. The possibility of an isotropic unjacketed response is pointed out, which - if verified experimentally - will make for an interesting and practically useful special case of anisotropic poroelasticity.

1 Introduction

This chapter offers a comprehensive review of the constitutive equations of linear poroelasticity theory or Biot Theory, as it is sometimes named after its founder Biot (1941). Special attention is given to anisotropic materials and to alternative formulations in terms of *drained* and *undrained* coefficients, bringing out the well-known analogy (Biot, 1956a; Geertsma, 1957a; Rice and Cleary, 1976) with isothermal and adiabatic coefficients in linear thermoelasticity theory (see, for example, Nye (1957) or Weiner (1983)). The parallelism of these theories sheds light on the relations between various material parameters, for example that between storage capacities at constant stress and constant strain, which mirrors the relation between the heat capacities at constant stress and constant strain, respectively. In order to display this analogy in full, only isothermal conditions are considered in this section, and readers interested in non-isothermal poroelasticity are directed to a paper by McTigue (1986) and to the monograph by Coussy

(2004). The outcome of the present survey is a complete set of linearized constitutive relations, phrased alternatively in terms of undrained bulk parameters or drained skeleton parameters and displayed in the form of a mnemonic diagram. This affords a rapid overview that should assist in the formulation of experimental strategies for the measurement of poroelastic constants. The principal relationships between alternative sets of material parameters are also tabulated. Following a discussion of unjacketed tests, we consider the possibility of an isotropic unjacketed response of otherwise anisotropic materials, and in the subsequent review of the fully isotropic case we discuss and compare different parameter definitions. The last section focusses on Gassmann's well-known fluid substitution relation and its generalization, i.e., the relationship between the rank-4 constitutive coefficients of a drained and an undrained description, which may be expressed in terms the rank-2 and rank-0 constitutive coefficients. This review is based on an unpublished report written for Shell Research in 1997 as a detailed and explicit guide to experimental studies of anisotropic poroelastic materials. The subject of this review has of course been dealt with repeatedly in the past, beginning with a paper by Biot (1955) up to recent treatments such as, for example, by Thompson and Willis (1991) and by Cheng (1997). None of these, however, appeared to us quite as systematic and comprehensive as called for by the theory's inherent structure. In general, and in the interest of revealing this structure in a consistent and concise manner, we have therefore dealt with earlier work only in direct connection with a particular aspect or point of controversy.

2 Work Potentials for Isothermal Deformation

The linear theory of poroelasticity[1] was first proposed by Biot (1941) and apparently conceived primarily as a generalization of an already existing one-dimensional theory of soil consolidation by Terzaghi and Fröhlich (1936). Biot's theory was modeled after the much older Duhamel-Neumann theory of thermoelasticity and its more recent thermodynamic foundation. This analogy has also suggested a proper point of departure for more general, nonlinear theories (Biot, 1972); see also Coussy (2004).

In keeping with this thermodynamic approach, we now focus, as in Rice and Cleary (1976) and in Rudnicki (1985, 2001), on a material element of a porous, fluid-saturated solid that occupies a volume V_0 in a chosen reference state. One may think of this element as being defined by the set of material points of the solid skeleton that constitute an imagined *cut*

[1]The term was coined by Geertsma (1966).

by which the volume element is separated from its surroundings. On the macroscopic scale of the porous body under consideration, this volume element is now viewed as an infinitesimal element that is subject to a locally homogeneous deformation, in the course of which it is brought from the reference state to the *current* state. Here we shall consider only isothermal changes of state in which the solid skeleton of the porous solid experiences but infinitesimal elastic deformations, such that macroscopic displacement gradients remain numerically much smaller than one. If V be the bulk volume of the chosen material volume element in the current state, V_ϕ will denote the pore volume, i.e., the part of V taken up by the interconnected pore space; then $\phi = V_\phi/V$ represents the porosity of the material element. And if p_0 and p be the pressures and $\rho_0(p_0)$ and $\rho(p)$ the mass densities of a homogeneous pore fluid in the reference and current state, respectively, then $m = \rho V_\phi/V_0 = \rho J\phi$ denotes the current fluid mass per unit bulk volume in the reference configuration and is therefore a referential partial mass density of the pore fluid, $J = V/V_0$ being the Jacobian of the deformation from the reference to the current state (see, e.g., Coussy (2004), Chap. I). In the conceptual frame of thermodynamics, the surroundings of our infinitesimal volume element are treated as a 'reservoir' with which it can exchange pore fluid *reversibly* in the course of which the fluid pressure will perform positive or negative work on the infinitesimal material element. An increment of this *hydraulic work*, as one may call it, can be expressed as the product $\mu V_0 \mathrm{d}m$ of an incremental fluid mass and the *pressure function*

$$\mu = \int_{p_0}^{p} \frac{\mathrm{d}p}{\rho}, \tag{1}$$

whose reference value at p_0 is set equal to zero. To see this, we first note that

$$\int_{p_0}^{p} \frac{\mathrm{d}p}{\rho} = -\frac{p_0}{\rho_0} - \int_{p_0}^{p} p\,\mathrm{d}\left(\frac{1}{\rho}\right) + \frac{p}{\rho}.$$

This provides an interpretation of μ as the *isothermal* work done on a unit mass of fluid in three consecutive steps, which together amount to a change from the *reference state* (p_0, ρ_0) to the *current state* (p, ρ) of the fluid. In thermodynamics, this *reversible* change is pictured in terms of two very large fluid reservoirs that are maintained in these two states, and a small cylinder equipped with a frictionless piston that is used to (a) extract a unit mass of fluid at the constant pressure p_0 from the first reservoir by performing the (negative) work $-p_0 \int_0^{1/\rho_0} \mathrm{d}(1/\rho) = -p_0/\rho_0$ on the fluid, (b) compress the fluid in the cylinder to the current state (p, ρ) of the second reservoir by performing the (positive) work $-\int_{p_0}^{p} p\,\mathrm{d}(1/\rho)$, and (c) to inject the fluid at the constant pressure p into the second reservoir by performing the (positive) work $-p\int_{1/\rho}^{0} \mathrm{d}(1/\rho) = p/\rho$. As shown by Biot (1972), the product $\mu\,\mathrm{d}m$ therefore represents the work performed per unit volume in *transferring* a differential fluid mass across the imagined *material* boundary of a volume element, the quantity μ playing the role of a *chemical potential* of the percolating pore fluid.

If, on the other hand, the volume element is imagined to be sealed or kept undrained at constant m, the increment of reversible work absorbed or given out by the element must be given - as with an ordinary elastic material - by the product $V_0\sigma_{ij}\mathrm{d}\varepsilon_{ij}$ of the increment $\mathrm{d}\varepsilon_{ij}$ in infinitesimal strain and the total (Cauchy) stress σ_{ij}, the latter being defined so that the force per unit area of an imagined cut through the porous medium is $n_i\sigma_{ij}$ where n_i are the components of the local unit normal of this cut.

In bringing the element from a given reference state to some current state, the total work performed per unit reference volume during a reversible change is now required to depend solely on the final state. This implies the existence of an isothermal work potential $\psi = \psi(\varepsilon_{ij}, m)$ such that any infinitesimal change

$$\mathrm{d}\psi = \sigma_{ij}\mathrm{d}\varepsilon_{ij} + \mu\,\mathrm{d}m \tag{2}$$

becomes an exact differential; the function $\psi(\varepsilon_{ij}, m)$ is thus seen to be identical with a Helmholtz free energy per unit reference volume.

Next we consider the Helmholtz free energy density of the pore fluid ψ_f per unit bulk volume in the reference state, which is defined in terms of the Gibbs free energy density $\rho\mu$, the pressure p and the pore volume fraction $v = V_\phi/V_0 = \phi J$ by

$$\psi_f = v(\rho\mu - p). \tag{3}$$

We are assuming here that the fluid mass content m depends solely on the fluid density ρ and the pore volume fraction v; the increment of the latter is therefore given by

$$\mathrm{d}v = \mathrm{d}\left(\frac{m}{\rho}\right). \tag{4}$$

The free energy ψ_f may now be subtracted from ψ to form the free energy $\tilde{\psi} = \psi - \psi_f$, whose differential is found to satisfy the Gibbs equation[2]

[2]M. Biot, in 1941, developed poroelasticity theory from this potential, writing $U(\varepsilon_{kl}, \theta)$ for $\tilde{\psi}(\varepsilon_{kl}, v)$. Guéguen *et al.* (2004) obtain (5) along an alternative route from a pore-scale description, using a spatial averaging technique.

$$\mathrm{d}\tilde{\psi} = \sigma_{ij}\,\mathrm{d}\varepsilon_{ij} + p\,\mathrm{d}v, \tag{5}$$

as will be evident from (2)–(4).

As Biot has observed, $\tilde{\psi}$ can be interpreted as the free energy density of a *wetted solid* composed of a solid matrix and a thin layer of fluid attached to the pore walls, i.e., forming part of the porous skeleton. Relation (4) then demands that we assume the mass of this fluid layer to remain constant. The attached fluid layer may also be viewed as the site of any interfacial interactions between the solid and the pore fluid. The free energy ψ is thereby interpreted as the sum of the free energies of two phases that interact only mechanically through the work $p\mathrm{d}v$ performed by the pore fluid pressure upon a wetted solid skeleton. We can now construct a constitutive theory of (isothermal) poroelasticity from the fundamental equations $\psi = \psi(\varepsilon_{ij}, m)$ and $\tilde{\psi} = \tilde{\psi}(\varepsilon_{ij}, v)$ as follows.

We subtract the total differential of the potential $\psi = \psi(\varepsilon_{ij}, m)$ from the Gibbs equation (2) to get

$$\left(\sigma_{ij} - \frac{\partial\psi}{\partial\varepsilon_{ij}}\right)\mathrm{d}\varepsilon_{ij} + \left(\mu - \frac{\partial\psi}{\partial m}\right)\mathrm{d}m = 0,$$

and observe that this must hold for arbitrary differentials $\mathrm{d}\sigma_{ij}$ and $\mathrm{d}m$. It therefore follows that the coefficients of these differentials must vanish and this implies the existence of the equations of state[3]

$$\sigma_{ij} = \left.\frac{\partial\psi}{\partial\varepsilon_{ij}}\right|_m \equiv \sigma_{ij}(\varepsilon_{kl}, m), \tag{6}$$

$$\mu = \left.\frac{\partial\psi}{\partial m}\right|_{\sigma_{ij}} \equiv \mu(\varepsilon_{ij}, m). \tag{7}$$

These equations may in general exhibit some *material* nonlinearity and they are understood to depend, just as the potential function $\psi(\varepsilon_{ij}, m)$, on the choice of reference configuration with respect to which ε_{ij} and m are measured.

[3]Here, and in subsequent analogous situations, it is assumed that the potential function has been symmetrized in its arguments ε_{ij} and ε_{ji} by replacing each of them by $(\varepsilon_{ij} + \varepsilon_{ji})/2$; it is thus regarded, for the purpose of partial differentiation, as a function of nine variables rather than of the six independent components of the symmetric strain tensor.

Similarly, for the potential $\tilde{\psi} = \tilde{\psi}(\varepsilon_{ij}, v)$ of (5) we obtain

$$\sigma_{ij} = \left.\frac{\partial \tilde{\psi}}{\partial \varepsilon_{ij}}\right|_v \equiv \sigma_{ij}(\varepsilon_{kl}, v), \tag{8}$$

$$p = \left.\frac{\partial \tilde{\psi}}{\partial v}\right|_{\sigma_{ij}} \equiv p(\varepsilon_{ij}, v). \tag{9}$$

Practical considerations will often suggest the use of different sets of independent variables. For example, we may wish choose the pore fluid pressure p instead of the volume fraction v for the obvious reason that it is the pressure rather than the pore volume fraction that is readily measured in the field or controlled in a laboratory experiment. This change of independent variables is achieved by means of a contact transformation or Legendre transformation (see, e.g., Callen (1960)). Accordingly, starting from the potential $\tilde{\psi}(\varepsilon_{ij}, v)$, we define the function

$$\tilde{\pi} = \tilde{\psi} - pv. \tag{10}$$

Taking its differential and using (5) we obtain

$$\mathrm{d}\tilde{\pi} = \sigma_{ij}\, \mathrm{d}\varepsilon_{ij} - v\, \mathrm{d}p \tag{11}$$

for the function $\tilde{\pi}(\varepsilon_{ij}, p)$ of the independent variables ε_{ij} and p. Thus we deduce for this potential the pair of equations of state

$$\sigma_{ij} = \left.\frac{\partial \tilde{\pi}}{\partial \varepsilon_{ij}}\right|_p \equiv \sigma_{ij}(\varepsilon_{kl}, p), \tag{12}$$

$$v = -\left.\frac{\partial \tilde{\pi}}{\partial p}\right|_{\varepsilon_{ij}} \equiv v(\varepsilon_{ij}, p). \tag{13}$$

Since the variables $\sigma_{ij}, \varepsilon_{ij}, m, \mu, v, p$ can be grouped in eight different pairs of independent variables, each comprising one scalar and one tensorial quantity, one can define eight different thermodynamic potentials. These are listed in Table 1, each with its corresponding pair of state equations.

A word must be said here about the notation used in Table 1. Note, first, that there are two sets of scalar independent variables–m, v and μ, p–that can characterize two distinct modes of loading and associated mechanical response of a porous, fluid-saturated material. The distinction that one draws here is between an *undrained* state of a material volume element of porous body, in which no fluid mass is lost or gained by the element while it is being deformed, and a *drained* state in which the fluid mass is allowed to be stored in or drained from the element at a constant fluid pressure.

Table 1. Thermodynamic potentials and equations of state in isothermal poroelasticity

drained description		undrained description	
$\pi(\varepsilon_{kl},\mu)$:	$\sigma_{ij}(\varepsilon_{kl},\mu)=\frac{\partial\pi}{\partial\varepsilon_{ij}}\vert_{\mu}$ $m(\varepsilon_{kl},\mu)=\frac{-\partial\pi}{\partial\mu}\vert_{\varepsilon_{kl}}$	$\psi(\varepsilon_{kl},m)$:	$\sigma_{ij}(\varepsilon_{kl},m)=\frac{\partial\psi}{\partial\varepsilon_{ij}}\vert_{m}$ $\mu(\varepsilon_{kl},m)=\frac{\partial\psi}{\partial m}\vert_{\varepsilon_{kl}}$
$\chi(\sigma_{kl},\mu)$:	$\varepsilon_{ij}(\sigma_{kl},\mu)=\frac{-\partial\chi}{\partial\sigma_{ij}}\vert_{\mu}$ $m(\sigma_{kl},\mu)=\frac{-\partial\chi}{\partial\mu}\vert_{\sigma_{kl}}$	$\varphi(\sigma_{kl},m)$:	$\varepsilon_{ij}(\sigma_{kl},m)=\frac{-\partial\varphi}{\partial\sigma_{ij}}\vert_{m}$ $\mu(\sigma_{kl},m)=\frac{\partial\varphi}{\partial m}\vert_{\sigma_{kl}}$
$\tilde{\pi}(\varepsilon_{kl},p)$:	$\sigma_{ij}(\varepsilon_{kl},p)=\frac{\partial\tilde{\pi}}{\partial\varepsilon_{ij}}\vert_{p}$ $v(\varepsilon_{kl},p)=\frac{-\partial\tilde{\pi}}{\partial p}\vert_{\varepsilon_{kl}}$	$\tilde{\psi}(\varepsilon_{kl},v)$:	$\sigma_{ij}(\varepsilon_{kl},v)=\frac{\partial\tilde{\psi}}{\partial\varepsilon_{ij}}\vert_{v}$ $p(\varepsilon_{kl},v)=\frac{\partial\tilde{\psi}}{\partial v}\vert_{\varepsilon_{kl}}$
$\tilde{\chi}(\sigma_{kl},p)$:	$\varepsilon_{ij}(\sigma_{kl},p)=\frac{-\partial\tilde{\chi}}{\partial\sigma_{ij}}\vert_{p}$ $v(\sigma_{kl},p)=\frac{-\partial\tilde{\chi}}{\partial p}\vert_{\sigma_{kl}}$	$\tilde{\varphi}(\sigma_{kl},v)$:	$\varepsilon_{ij}(\sigma_{kl},v)=\frac{-\partial\tilde{\varphi}}{\partial\sigma_{ij}}\vert_{v}$ $p(\sigma_{kl},v)=\frac{\partial\tilde{\varphi}}{\partial v}\vert_{\sigma_{kl}}$

Since, under isothermal conditions, μ is uniquely determined by the pressure p, a constitutive formulation employing either p or μ as independent scalar variable will be referred to a *drained description*, which is to say: a description in terms of parameters obtained from experiments that were performed under drained conditions. On the other hand, when the fluid mass content m is held fixed in an experiment during loading, the sample will exhibit an *undrained* response and this will reflect the response of both the solid and the fluid phase. Accordingly, we shall speak of an *undrained description*, if it is phrased in terms of parameters that were obtained from experiments performed under undrained conditions in which either m or v were held fixed. The case $v = constant$ is somewhat special. Here the contemplated way of controlling the pore volume fraction in an experiment, at least in principle, is to employ an incompressible pore fluid. For this reason the choice of v as an independent variable corresponds to a special case of an undrained description.

Secondly, we shall adopt a notation that distinguishes between the thermodynamic potentials of a fluid-saturated bulk volume element, and potentials that are unaffected by properties of the mobile pore fluid, i.e., characterize only the wetted solid skeleton of Biot. An example of the latter is the potential $\tilde{\psi}(\varepsilon_{ij}, v)$ and the same superimposed tilde is used in Table 1 to denote the other three potentials for the wetted solid. Note, therefore, that the linear constitutive theories constructed from the four potentials listed in the upper half of Table 1 presuppose a constant compressibility of the pore fluid and for this reason will have a more restricted range of applicability than their counterparts that derive from the potentials of the lower half of the table and which contain no reference to the equations of state of the pore fluid. This difference can become important, for example in dealing with a highly compressible pore fluid, such as natural gas, but an approximately linear elastic skeleton response over substantial pressure differentials. A drained description based on the potentials $\tilde{\pi}(\varepsilon_{kl}, p)$ or $\tilde{\chi}(\sigma_{kl}, p)$ will be the appropriate choice in such cases.

3 Linear Constitutive Relations

We now consider the various linear constitutive relations that may be constructed as approximations of the equations of state listed in Table 1. This discussion is divided according to the potential considered into two subsections, dealing with formulations in terms of undrained and drained properties, respectively. The reader will no doubt observe and perhaps be tired by a certain repetitiveness in what follows. Here the writer's intention was to expose the inherent structural symmetries of the theory in a way that can be summarized succinctly in a final table and diagram, the individual entries of which should nevertheless remain easy to trace back to their sources in a sufficiently organized text. In the end, Table 2 and Figure 1 could thus become the only things to be looked at as the occasion arises.

3.1 Undrained Description

THE POTENTIAL $\psi(\varepsilon_{kl}, m)$. The formal linearization procedure that we shall adopt here is to truncate a Taylor series expansion about a given reference state of a set of state equations derived from a chosen thermodynamic potential. Accordingly, in linearizing the state equations (6),(7), we specify the reference configuration of the porous solid under consideration as one in which the strain ε_{ij} vanishes throughout the body and the pore fluid has the density $\rho_0(x_1, x_2, x_3)$, the volume fraction m/ρ therefore being equal to v_0, and by requiring that when the body is in this state, equations (6) and (7) satisfy the relations

$$\sigma^0_{ij} = \sigma_{ij}(0, \rho_0 v_0) \qquad \text{and} \qquad 0 = \mu(0, \rho_0 v_0).$$

Here the *initial stress* σ^0_{ij} is understood to form an equilibrium system of stresses in a gravitational field, satisfying the equations of equilibrium $\partial\sigma^0_{ij}/\partial x_j + \overline{\rho}_0 g_i = 0$, where $\overline{\rho}_0$ denotes the bulk density of the (fluid-saturated) porous solid in the reference state and g_i the component of the acceleration of gravity in the direction of the coordinate x_i (only rectangular Cartesian coordinates are considered).

In expanding (6) & (1), we continue to assume small displacement gradients. In addition, so as to obtain a fully linear undrained description, we shall permit only such changes in fluid mass content as will be compatible with the assumption of a constant fluid compressibility. We shall see presently, however, that linear poroelasticity is not contingent upon this last assumption.

The linear approximations of (6) & (1) thus become

$$\sigma_{ij}(\varepsilon_{kl}, m_0 + \Delta m; R) = \sigma_{ij}|^0 + \left.\frac{\partial\sigma_{ij}}{\partial\varepsilon_{kl}}\right|^0_m \varepsilon_{kl} + \left.\frac{\partial\sigma_{ij}}{\partial m}\right|^0_{\varepsilon_{kl}} \Delta m$$

and

$$\mu(\varepsilon_{ij}, m_0 + \Delta m) = \mu|^0 + \left.\frac{\partial\mu}{\partial\varepsilon_{ij}}\right|^0_m \varepsilon_{ij} + \left.\frac{\partial\mu}{\partial m}\right|^0_{\varepsilon_{ij}} \Delta m,$$

where the superscript '0' denotes evaluation at the reference state, $\sigma_{ij}|^0 = \sigma^0_{ij}$ denotes the *initial stress* as specified above, and $\mu|^0 = 0$. We now express these linear relationships in the form

$$\sigma_{ij} = \sigma^0_{ij} + C^u_{ijkl}\,\varepsilon_{kl} - a_{ij}\Delta m/\rho_0, \tag{14}$$

and

$$\mu = -\rho_0^{-1} a_{ij}\,\varepsilon_{ij} + \rho_0^{-1} M_\varepsilon \Delta m/\rho_0, \tag{15}$$

by defining the following constant coefficients: The rank-4 tensor of *undrained elastic constants* or *undrained stiffnesses* with components

$$C^u_{ijkl} = \left.\frac{\partial \sigma_{ij}}{\partial \varepsilon_{kl}}\right|^0_m = \left.\frac{\partial^2 \psi}{\partial \varepsilon_{kl} \partial \varepsilon_{ij}}\right|_0, \tag{16}$$

a rank-2 tensor with components

$$a_{ij} = -\rho_0 \left.\frac{\partial \sigma_{ij}}{\partial m}\right|^0_{\varepsilon_{kl}} = -\rho_0 \left.\frac{\partial^2 \psi}{\partial m \partial \varepsilon_{ij}}\right|_0 = -\rho_0 \left.\frac{\partial \mu}{\partial \varepsilon_{ij}}\right|^0_m, \tag{17}$$

expressing the change in stress with fluid mass content when the strain is held fixed and equal to zero (or the change in the pressure function with strain when the mass content is held fixed), and the scalar

$$\rho_0^{-1} M_\varepsilon = \rho_0 \left.\frac{\partial \mu}{\partial m}\right|^0_{\varepsilon_{ij}} = \rho_0 \left.\frac{\partial^2 \psi}{\partial m^2}\right|_0. \tag{18}$$

Note the appearance of the same coefficient a_{ij} in both (14) and (15), which is implied by a Maxwell-type reciprocity relation that derives from the interchangeability of the order of differentiation in the mixed derivatives of the potential ψ in (17).

It is not difficult to see, that a change in fluid mass content at constant strain will, in general, evoke a non-hydrostatic stress response and that this demands a tensorial coefficient a_{ij}. One only has to think of a porosity that is at least partly due to the presence of low-aspect ratio, crack-like pores that display a certain measure of preferred orientation.

A straightforward interpretation of the coefficient M_ε will be given further below in terms of its reciprocal $1/M_\varepsilon$, which has the significance of a specific storage capacity at constant strain. We note here that M_ε is identical with the modulus M of Biot and Willis (1957). In the present context the subscript 'ε' serves to distinguish M_ε from a similar modulus M_σ (cf. Eq. 26) that is defined at constant stress rather than strain.

The symmetry of ε_{ij} and σ_{ij} implies the symmetries of a_{ij} and C^u_{ijkl} with respect to the interchanges $i \leftrightarrow j$ and $k \leftrightarrow l$ of indices, and a further symmetry of C^u_{ijkl} with respect to the pairwise interchange $ij \leftrightarrow kl$ follows from the interchangeability in (16) of the order of differentiation of the potential function $\psi(\varepsilon_{kl}, m)$ with respect to the strain components. The symmetries

$$a_{ij} = a_{ji} \quad \text{and} \quad C^u_{ijkl} = C^u_{jikl} = C^u_{ijlk} = C^u_{klij} \tag{19}$$

must therefore apply, so that there are at most 6 independent components of a_{ij} and 21 independent components of C^u_{ijkl}. These numbers will be reduced for any a particular poroelastic material by further, material symmetries.

THE POTENTIAL $\varphi(\sigma_{kl}, m)$. Consider the set of state equations

$$\varepsilon_{ij} = -\frac{\partial \varphi}{\partial \sigma_{ij}}\Big|_m \equiv \varepsilon_{ij}(\sigma_{kl}, m) \tag{20}$$

$$\mu = \frac{\partial \varphi}{\partial m}\Big|_{\varepsilon_{ij}} \equiv \mu(\sigma_{kl}, m) \tag{21}$$

that derives from the potential $\varphi(\sigma_{kl}, m) = \psi(\varepsilon_{kl}, m) - \sigma_{ij}\varepsilon_{ij}$. We linearize these equations about a reference state in which $\sigma_{kl} = \sigma^0_{kl}$ and $m = \rho_0 v_0$, and in which (22) & (21) satisfy the conditions

$$0 = \varepsilon_{ij}(\sigma^0_{kl}, \rho_0 v_0) \qquad \text{and} \qquad \mu_0 = \mu(\sigma^0_{kl}, \rho_0 v_0).$$

With $\Delta m = m - \rho_0 v_0$ the resulting linear constitutive relations are

$$\varepsilon_{ij} = S^u_{ijkl}\,(\sigma_{kl} - \sigma^0_{kl}) + b_{ij}\,\Delta m/\rho_0, \tag{22}$$

and

$$\mu = -\rho_0^{-1} b_{ij}\,(\sigma_{ij} - \sigma^0_{ij}) + \rho_0^{-1} M_\sigma \Delta m/\rho_0. \tag{23}$$

The constant coefficients in (22) and (23) are the components of the rank-4 tensor of *undrained elastic compliances*

$$S^u_{ijkl} = \frac{\partial \varepsilon_{ij}}{\partial \sigma_{kl}}\Big|^0_m = -\frac{\partial^2 \varphi}{\partial \sigma_{kl}\partial \sigma_{ij}}\Big|_0, \tag{24}$$

the components of a symmetric rank-2 tensor

$$b_{ij} = \rho_0 \frac{\partial \varepsilon_{ij}}{\partial m}\Big|^0_{\sigma_{kl}} = -\rho_0 \frac{\partial^2 \varphi}{\partial m \partial \sigma_{ij}}\Big|_0 = -\rho_0 \frac{\partial \mu}{\partial \sigma_{ij}}\Big|_m, \tag{25}$$

expressing the change in strain with fluid mass content when the stresses are held fixed at their initial values (or the change in the pressure function with stress when the mass content is held fixed), and the scalar

$$\rho_0^{-1} M_\sigma = \rho_0 \frac{\partial \mu}{\partial m}\Big|^0_{\sigma_{ij}} = \rho_0 \frac{\partial^2 \varphi}{\partial m^2}\Big|_0, \tag{26}$$

which we shall interpret presently through its reciprocal $1/M_\sigma$.

We note here that coefficient defined by (25) is the same as Thompson and Willis (1991) coefficient b_{ij}; it represents, as we shall see, an appropriate generalization of Skempton's scalar pore pressure coefficient B, such that $b_{ij} = b\delta_{ij} = \frac{1}{3}B\delta_{ij}$ for an isotropic medium. The appearance of this coefficient in both (22) and (23) is implied by a Maxwell-type reciprocity relation that derives from the interchangeability of the order of differentiation in the mixed derivatives of the potential φ in (25).

Two comments are now in order. The first applies to each of the various renderings in this section of the linear theory. It is the observation that the coefficients $C^u_{ijkl}, a_{ij}, M_\varepsilon$ and $S^u_{ijkl}, b_{ij}, M_\sigma$ may be interpreted as specifying tangent directions in ε_{ij}, m- respectively σ_{ij}, m-space at the reference state, on which they depend. The linearization procedure we have just described is therefore general enough to allow for an *incrementally linear* description near *any* suitably selected reference state.

Secondly, C^u_{ijkl} and S^u_{ijkl} were referred to as *undrained coefficients*, because they are defined in terms of partial derivatives taken at fixed fluid mass content. This terminology does *not* imply, however, that the use of these constants is restricted to conditions of constant fluid mass content, although it will be a natural choice in such circumstances.[4]

THE POTENTIAL $\tilde{\psi}(\varepsilon_{kl}, v)$. The choice of this potential, which satisfies the Gibbs equation (5), is primarily of theoretical interest, since the state equations (8) & (9) are linearized in this case in the displacement gradient and in the change in volume fraction, i.e., in the two kinematic variables. Although the volume fraction v remains an impractical choice of independent variable, the condition $|\Delta v| = |v - v_0| \ll 1$ represents nonetheless the most appropriate among the constraints that may be imposed on the various possible scalar independent variables. For while it is consistent with the idea of infinitesimal elastic deformations of the solid skeleton material of a porous body, it imposes no direct constraint on the magnitude of pore pressure changes. This also explains why Biot (1941) chose to develop his theory from the potential $\tilde{\psi}(\varepsilon_{kl}, v)$.

A reference configuration is now defined for the porous body under consideration such that the strain ε_{ij} vanishes and that its pore volume fraction equals $v_0(x_1, x_2, x_3)$ in this configuration, while equations (8) and (9) satisfy

$$\sigma^0_{ij} = \sigma_{ij}(0, v_0) \qquad \text{and} \qquad p_0 = p(0, v_0).$$

A truncated Taylor series expansion of (8) & (9) now yields the linear

[4] An undrained response may be expected, for example, for the propagation of low-frequency (ω) seismic waves, where the relaxation time for pore pressure diffusion (which is proportional to the square of the distance between rarefied and compressed regions over which fluid transport must take place) is large in comparison with $1/\omega$.

constitutive relations

$$\sigma_{ij} = \sigma_{ij}^0 + \tilde{C}_{ijkl}^u \varepsilon_{kl} - \tilde{a}_{ij}\Delta v. \tag{27}$$

and

$$p = p_0 - \tilde{a}_{ij}\,\varepsilon_{ij} + \tilde{M}_\varepsilon \Delta v, \tag{28}$$

Here

$$\tilde{C}_{ijkl}^u = \left.\frac{\partial \sigma_{ij}}{\partial \varepsilon_{kl}}\right|_v^0 = \left.\frac{\partial^2 \tilde{\psi}}{\partial \varepsilon_{kl} \partial \varepsilon_{ij}}\right|_0 \tag{29}$$

are the components of a rank-4 tensor of undrained elastic constants or stiffnesses, pertaining to the special case of an incompressible pore fluid,

$$\tilde{a}_{ij} = -\left.\frac{\partial \sigma_{ij}}{\partial v}\right|_{\varepsilon_{kl}}^0 = -\left.\frac{\partial^2 \tilde{\psi}}{\partial v \partial \varepsilon_{ij}}\right|_0 = -\left.\frac{\partial p}{\partial \varepsilon_{ij}}\right|_v^0 \tag{30}$$

are the components of a symmetric rank-2 tensor that expresses the change in stress with the pore volume fraction when the strain is held fixed and equal to zero (or the change in pore pressure when the pore volume fraction–or incompressible fluid mass content–is held fixed), and

$$\tilde{M}_\varepsilon = \left.\frac{\partial p}{\partial v}\right|_{\varepsilon_{ij}}^0 = \left.\frac{\partial^2 \tilde{\psi}}{\partial v^2}\right|_0. \tag{31}$$

Note again the appearance of the same coefficient $\tilde{a}_{ij}$ in both (27) and (28), which is implied by a Maxwell-type reciprocity relation that derives from the interchangeability of the order of differentiation in the mixed derivatives of the potential $\tilde{\psi}$ in (30). This enables us to identify the volume fraction coefficient in (27) with the strain coefficient in (28) and allows us to interpret $\tilde{S}_{ijkl}^u$ and $\tilde{a}_{ij}$ as coefficients pertaining to an undrained description, albeit one of a special kind for which the fluid density remains constant and equal to ρ_0.

THE POTENTIAL $\tilde{\varphi}(\sigma_{kl}, v)$. As a last set of state equations providing an undrained description, we consider the pair

$$\varepsilon_{ij} = -\left.\frac{\partial \tilde{\varphi}}{\partial \sigma_{ij}}\right|_v^0 \equiv \varepsilon_{ij}(\sigma_{kl}, v), \tag{32}$$

$$p = \left.\frac{\partial \tilde{\varphi}}{\partial v}\right|_{\varepsilon_{ij}}^0 \equiv p(\sigma_{kl}, v), \tag{33}$$

which can be derived from the potential $\tilde{\varphi}(\sigma_{kl}, v) = \tilde{\psi}(\varepsilon_{kl}, v) - \sigma_{ij}\varepsilon_{ij}$. Linearizing about a reference state in which $\sigma_{kl} = \sigma_{kl}^0$ and $v = v_0$, and in which (32) & (33) satisfy the conditions

$$0 = \varepsilon_{ij}(\sigma_{kl}^0, v_0) \qquad \text{and} \qquad p_0 = p(\sigma_{kl}^0, v_0),$$

we obtain

$$\varepsilon_{ij} = \tilde{S}^u_{ijkl}\,(\sigma_{kl} - \sigma^0_{kl}) + \tilde{b}_{ij}\,\Delta v, \tag{34}$$

and

$$p = p_0 - \tilde{b}_{ij}\,(\sigma_{ij} - \sigma^0_{ij}) + \tilde{M}_\sigma \Delta v. \tag{35}$$

Here

$$\tilde{S}^u_{ijkl} = \left.\frac{\partial \varepsilon_{ij}}{\partial \sigma_{kl}}\right|^0_v = -\left.\frac{\partial^2 \tilde{\varphi}}{\partial \sigma_{kl}\partial \sigma_{ij}}\right|_0 \tag{36}$$

denote the components of the tensor of *undrained elastic compliances*,

$$\tilde{b}_{ij} = \left.\frac{\partial \varepsilon_{ij}}{\partial v}\right|^0_{\sigma_{kl}} = -\left.\frac{\partial^2 \tilde{\varphi}}{\partial v \partial \sigma_{ij}}\right|_0 = -\left.\frac{\partial p}{\partial \sigma_{ij}}\right|^0_v \tag{37}$$

expresses the change in strain with the pore volume fraction when the stresses are held fixed at their initial values (or the change in pore pressure with stress when the volume fraction is held fixed it its initial value), and

$$\tilde{M}_\sigma = \left.\frac{\partial p}{\partial v}\right|^0_{\sigma_{ij}} = \left.\frac{\partial^2 \tilde{\varphi}}{\partial v^2}\right|_0. \tag{38}$$

The appearance of the same coefficient $\tilde{b}_{ij}$ in both (34) and (35) is implied by the reciprocity relation expressed in (37). This enables us to identify the volume fraction coefficient in (27) with the strain coefficient in (28) and allows us to interpret both, $\tilde{C}^u_{ijkl}$ and $\tilde{b}_{ij}$ as coefficients pertaining to a special undrained description in which the pore fluid is assumed to be incompressible.

Relationships between undrained coefficients. From (16) and (24) it follows by an application of the chain rule that

$$S^u_{ijkl}\,C^u_{klmn} = \left.\frac{\partial \varepsilon_{ij}}{\partial \sigma_{kl}}\right|^0_m \left.\frac{\partial \sigma_{kl}}{\partial \varepsilon_{mn}}\right|^0_m = \frac{1}{2}\,(\delta_{im}\delta_{jn} + \delta_{in}\delta_{jm})\,. \tag{39}$$

Making use of this result in (14), after multiplication by S^u_{ijkl}, we get

$$S^u_{ijkl}\,C^u_{klmn}\,\varepsilon_{mn} = \varepsilon_{ij} = S^u_{ijkl}(\sigma_{kl} - \sigma^0_{kl}) + S^u_{ijkl}\,a_{kl}\Delta m/\rho_0$$

and, upon comparing this with (22),

$$b_{ij} = S^u_{ijkl}\,a_{kl}. \tag{40}$$

Similarly, one finds that

$$C^u_{ijkl}\,S^u_{klmn} = \frac{1}{2}\,(\delta_{im}\delta_{jn} + \delta_{in}\delta_{jm})\,. \tag{41}$$

Using this property in (22), after multiplication by C^u_{ijkl}, gives

$$\sigma_{ij} - \sigma^0_{ij} = C^u_{ijkl}\,\varepsilon_{kl} - C^u_{ijkl}\,b_{kl}\Delta m/\rho_0$$

and, upon comparing this with (14),

$$a_{ij} = C^u_{ijkl}\,b_{kl}. \tag{42}$$

Further, it is seen from (15) and (23) that

$$(M_\sigma - M_\varepsilon)\Delta m/\rho_0 = b_{ij}(\sigma_{ij} - \sigma^0_{ij}) - a_{ij}\varepsilon_{ij},$$

hence it follows from (14), (40) and (42) that

$$M_\varepsilon - M_\sigma = a_{ij}\,b_{ij} = a_{ij}\,S^u_{ijkl}\,a_{kl} = b_{ij}\,C^u_{ijkl}\,b_{kl}. \tag{43}$$

The same relationships evidently exist among the coefficients that are distinguished by a tilde as pertaining to an undrained description for incompressible pore fluids.

3.2 Drained Description

THE POTENTIAL $\tilde{\pi}(\varepsilon_{kl}, p)$. The equations of state (12) & (13) associated with this potential are linearized about a reference state in which the strain ε_{ij} vanishes and by demanding that they satisfy

$$\sigma^0_{ij} = \sigma_{ij}(0, p_0) \quad \text{and} \quad v_0 = v(0, p_0).$$

when the pore pressure equals $p_0(x_1, x_2, x_3)$ throughout a given porous body.

In truncating the series expansion (12) & (13) about this reference state after the linear terms, we shall as usual assume all components of displacement gradient to remain much smaller than one. Note, however, that we need not impose the same constraint on the relative change in pore pressure. The only constraint that we impose on pore pressure changes is that they remain consistent with the kinematic assumption of small changes in the pore volume fraction. The resulting linear constitutive relations are now written

$$\sigma_{ij} = \sigma^0_{ij} + C_{ijkl}\,\varepsilon_{kl} - \alpha_{ij}\Delta p \tag{44}$$

and

$$v = v_0 + \alpha_{ij}\,\varepsilon_{ij} + \tilde{S}_\varepsilon \Delta p, \tag{45}$$

with the *drained* stiffnesses

$$C_{ijkl} = \left.\frac{\partial \sigma_{ij}}{\partial \varepsilon_{kl}}\right|^0_p = \left.\frac{\partial^2 \tilde{\pi}}{\partial \varepsilon_{kl}\partial \varepsilon_{ij}}\right|_0, \tag{46}$$

the *pore pressure coefficient* or *Biot coefficient*

$$\alpha_{ij} = -\frac{\partial \sigma_{ij}}{\partial p}\bigg|^0_{\varepsilon_{kl}} = -\frac{\partial^2 \tilde{\pi}}{\partial p \partial \varepsilon_{ij}}\bigg|_0 = \frac{\partial v}{\partial \varepsilon_{ij}}\bigg|^0_p, \tag{47}$$

and the *specific storage capacity* or *specific storage coefficient* at constant strain[5]

$$\tilde{S}_\varepsilon = \frac{\partial v}{\partial p}\bigg|^0_{\varepsilon_{ij}} = -\frac{\partial^2 \tilde{\pi}}{\partial p^2}\bigg|_0. \tag{48}$$

We recall here that in Biot (1941) the change Δv in the void volume fraction is denoted by θ; also, the coefficients α_{ij} and $\tilde{S}_\varepsilon$ correspond to Biot's $\alpha\delta_{ij}$ and Q^{-1}, respectively, for isotropic porous media.

An immediate consequence of (44) is that the quantity

$$\Delta\sigma'_{ij} = \sigma_{ij} - \sigma^0_{ij} + \alpha_{ij}\,\Delta p \tag{49}$$

furnishes an appropriate generalization for anisotropic poroelastic materials of the familiar Biot-Willis *effective stress* (Biot and Willis, 1957; Geertsma, 1957b; Nur and Byerlee, 1971). The existence of an *effective stress principle* in poroelasticity is in fact a necessary consequence of linearization alone and is contingent upon no additional requirements. It is independent, in particular, of micro-mechanical properties other than the elastic skeleton response implied by the above formulation based on work potentials, although it is well known (Nur and Byerlee, 1971; Carroll, 1979) that the coefficient α_{ij} is more directly related to the pore-scale elastic constants of the solid phase when these remain uniform throughout a representative elementary volume of the porous medium (see the discussion further below). However, independent of any micro-mechanical interpretation, the coefficient α_{ij} retains the straightforward significance given to it by (47) of a quantity that can be determined directly from suitable macroscopic experiments.

We note further that since the potentials $\tilde{\pi}$ and $\tilde{\psi}$ are related by the contact transformation $\tilde{\pi} = \tilde{\psi} - pv$, the pair (44),(45) of linear constitutive relations could in fact have been obtained directly by rearranging (27),(28) so as to resemble (44),(45) in form, and by determining the coefficients $C_{ijkl}, \alpha_{ij}, \tilde{S}_\varepsilon$ in terms of the given coefficients $\tilde{C}^u_{ijkl}, \tilde{a}_{ij}, \tilde{M}_\varepsilon$. Thus, by rewriting (28) as

$$\Delta v = \frac{1}{\tilde{M}_\varepsilon}\tilde{a}_{kl}\varepsilon_{kl} + \frac{1}{\tilde{M}_\varepsilon}\Delta p$$

[5]We consider *storage capacity* more telling than the customary *storage coefficient*; it is also suggestive of the scalar nature of this quantity. The qualification 'specific' is commonly understood to mean 'per unit mass', but a specific storage capacity can obviously express a capacity to store so many *kg* per *kg* per unit pressure change or so many m^3 per m^3 per unit pressure change.

and substituting this expression for Δv in (27), we obtain

$$\sigma_{ij} = \sigma_{ij}^0 + (\tilde{C}_{ijkl}^u - \frac{1}{\tilde{M}_\varepsilon}\tilde{a}_{ij}\tilde{a}_{kl})\varepsilon_{kl} - \frac{1}{\tilde{M}_\varepsilon}\tilde{a}_{ij}\Delta p.$$

Therefore the following relationships between the coefficients of the formulations (44),(45) and (27),(28) must hold:

$$C_{ijkl} = \tilde{C}_{ijkl}^u - \frac{1}{\tilde{M}_\varepsilon}\tilde{a}_{ij}\tilde{a}_{kl}, \qquad \alpha_{ij} = \frac{1}{\tilde{M}_\varepsilon}\tilde{a}_{ij}, \qquad \tilde{S}_\varepsilon = \frac{1}{\tilde{M}_\varepsilon}. \tag{50}$$

These provide a first set of relationships between the coefficients of a drained and those of an undrained description, albeit for the special case of an incompressible pore fluid.

THE POTENTIAL $\tilde{\chi}(\sigma_{kl}, p)$. Observing that $\tilde{\chi}(\sigma_{kl}, p) = \tilde{\pi}(\varepsilon_{kl}, p) - \sigma_{ij}\varepsilon_{ij}$ is obtained by a contact transformation from the previous potential, we avoid further repetition and write down the linear constitutive relations

$$\varepsilon_{ij} = S_{ijkl}\,(\sigma_{kl} - \sigma_{kl}^0) + \beta_{ij}\,\Delta p \tag{51}$$

and

$$v = v_0 + \beta_{ij}\,(\sigma_{ij} - \sigma_{ij}^0) + \tilde{S}_\sigma \Delta p, \tag{52}$$

where

$$S_{ijkl} = \left.\frac{\partial \varepsilon_{ij}}{\partial \sigma_{kl}}\right|_p^0 = -\left.\frac{\partial^2 \tilde{\chi}}{\partial \sigma_{kl}\partial \sigma_{ij}}\right|_0 \tag{53}$$

are the *drained elastic compliances*,

$$\beta_{ij} = \left.\frac{\partial \varepsilon_{ij}}{\partial p}\right|_{\sigma_{kl}}^0 = -\left.\frac{\partial^2 \tilde{\chi}}{\partial p\partial \sigma_{ij}}\right|_0 = \left.\frac{\partial v}{\partial \sigma_{ij}}\right|_p^0 \tag{54}$$

are the components of a symmetric *hydraulic expansion tensor*, and

$$\tilde{S}_\sigma = \left.\frac{\partial v}{\partial p}\right|_{\sigma_{ij}}^0 = -\left.\frac{\partial^2 \tilde{\chi}}{\partial p^2}\right|_0 \tag{55}$$

is a *hydraulic void expansion coefficient* at constant stress.

We note here that the parameter H^{-1} in Biot's 1941 paper represents the equivalent of the trace 3β of the tensor β_{ij} for the case of isotropy, while Biot's parameter R^{-1} is our $\tilde{S}_\sigma$.

From (52) it is apparent that the pore volume change is controlled by the *effective stress* quantity

$$\sigma_{ij} - \sigma_{ij}^0 + \tilde{S}_\sigma \beta_{ij}^{-1}\,\Delta p \tag{56}$$

where β_{ij}^{-1} is the inverse of β_{ij}.

The coefficients α_{ij}, C_{ijkl} and β_{ij}, S_{ijkl} are referred to as *drained* coefficients, because they are defined in terms of partial derivatives taken at fixed pore pressure. This terminology does *not* imply, however, that the use of these constants is restricted to conditions of constant pore pressure. Indeed, the full description of *any* poroelastic process may always be cast in terms of the variable sets σ_{ij}, p or ε_{ij}, p. The drained constants are of course the natural choice for the description of drained processes, i.e., poroelastic processes in which the pressure appears merely in the role of a constant parameter.

We note further that since the potentials $\tilde{\chi}$ and $\tilde{\varphi}$ are related by the contact transformation $\tilde{\chi} = \tilde{\varphi} - pv$, the pair (51),(52) of linear constitutive relations could in fact have been obtained directly by rearranging (34),(35) so as to resemble (51),(52) in form, and by determining the coefficients $S_{ijkl}, \beta_{ij}, \tilde{S}_\sigma$ in terms of the given coefficients $\tilde{S}^u_{ijkl}, \tilde{b}_{ij}, \tilde{M}_\sigma$. Thus, by rewriting (35) as

$$\Delta v = \frac{1}{\tilde{M}_\sigma}\tilde{b}_{kl}\varepsilon_{kl} + \frac{1}{\tilde{M}_\sigma}\Delta p$$

and substituting this expression for Δv in (34), we obtain

$$\sigma_{ij} = \sigma_{ij}^0 + (\tilde{S}^u_{ijkl} - \frac{1}{\tilde{M}_\sigma}\tilde{b}_{ij}\tilde{b}_{kl})\varepsilon_{kl} - \frac{1}{\tilde{M}_\sigma}\tilde{b}_{ij}\Delta p.$$

From these we deduce the following relationships between the coefficients of the formulations (51),(52) and (34),(35)

$$S_{ijkl} = \tilde{S}^u_{ijkl} - \frac{1}{\tilde{M}_\sigma}\tilde{b}_{ij}\tilde{b}_{kl}, \qquad \beta_{ij} = \frac{1}{\tilde{M}_\sigma}\tilde{b}_{ij}, \qquad \tilde{S}_\sigma = \frac{1}{\tilde{M}_\sigma}. \tag{57}$$

This represents a second set of relationships between the coefficients of a drained and those of an undrained description for the special case of an incompressible pore fluid.

THE POTENTIAL $\pi(\varepsilon_{kl}, \mu)$. An alternative drained description, based on the potential $\pi(\varepsilon_{kl}, \mu) = \psi(\varepsilon_{kl}, m) - \mu m$, is obtained directly upon rewriting the linear relationships (14),(15) in terms of the independent varible μ. Since the latter is a unique function of the pressure p, this will indeed correspond to another drained description[6]. Thus we write

$$\sigma_{ij} = \sigma_{ij}^0 + C_{ijkl}\varepsilon_{kl} - \alpha_{ij}\rho_0\mu, \tag{58}$$

[6]Note, however, that the present description will involve the assumption of a constant fluid compressibility, which is why the above linear drained descriptions based on the potentials $\tilde{\pi}(\varepsilon_{kl}, p)$ and $\tilde{\chi}(\sigma_{kl}, p)$ are the preferred ones for highly compressible pore fluids.

and

$$\Delta m/\rho_0 = \alpha_{ij}\,\varepsilon_{ij} + S_\varepsilon \rho_0 \mu, \tag{59}$$

observing that the stiffness tensor must be the same as in (44), and that

$$\begin{aligned}\alpha_{ij} &= -\frac{1}{\rho_0}\frac{\partial \sigma_{ij}}{\partial \mu}\bigg|^0_{\varepsilon_{kl}} = -\frac{1}{\rho_0}\frac{\partial^2 \pi}{\partial \mu \partial \varepsilon_{ij}}\bigg|_0 = \frac{1}{\rho_0}\frac{\partial m}{\partial \varepsilon_{ij}}\bigg|^0_{\mu} \\ &= -\frac{\partial \sigma_{ij}}{\partial \mu}\frac{\mathrm{d}\mu}{\mathrm{d}p}\bigg|^0_{\varepsilon_{kl}} = -\frac{\partial \sigma_{ij}}{\partial p}\bigg|^0_{\varepsilon_{kl}},\end{aligned}$$

the latter on account of (47). Further, in the linearised theory under consideration, the change in fluid mass content per unit reference volume may be written in terms of the changes in fluid density and void volume fraction as

$$\Delta m = m - m_0 = \rho_0(v - v_0) + (\rho - \rho_0)v_0, \tag{60}$$

so that one has

$$\frac{1}{\rho_0}\frac{\partial m}{\partial p}\bigg|^0_{\varepsilon_{kl}} = \frac{1}{\rho_0}\frac{\partial m}{\partial \mu}\frac{\mathrm{d}\mu}{\mathrm{d}p}\bigg|^0_{\varepsilon_{kl}} = \frac{\partial v}{\partial p}\bigg|^0_{\varepsilon_{kl}} + v_0\frac{1}{\rho_0}\frac{\partial \rho}{\partial p}\bigg|_0,$$

that is

$$S_\varepsilon = \tilde{S}_\varepsilon + v_0 c_f, \tag{61}$$

where $c_f = (1/\rho_0)(\partial\rho/\partial p)|_0 = (\rho - \rho_0)/\rho_0\Delta p$ is a constant compressibility of the fluid for the small changes in the fluid density implied by (60).

We return here for a moment to relation (45), adding a term $(\rho - \rho_0)v_0/\rho_0 = v_0 c_f \Delta p$ on both sides to note that the result can be written $\Delta m/\rho_0 = \alpha_{ij}\,\varepsilon_{ij} + (\tilde{S}_\varepsilon + v_0 c_f)\Delta p$. Evidently, therefore, the independent variable $\rho_0\mu$ may be replaced by the pore pressure change Δp in the above constitutive relations (58),(59), so that we recover (44) together with

$$\Delta m/\rho_0 = \alpha_{ij}\,\varepsilon_{ij} + S_\varepsilon \Delta p. \tag{62}$$

as an alternative set of constitutive relations. From this S_ε is seen to have the simple interpretation of a *specific storage capacity* at fixed strain and this also explains the nature of its reciprocal M_ε.

Putting now (15) into the form

$$\Delta m/\rho_0 = \frac{1}{M_\varepsilon} a_{kl}\,\varepsilon_{kl} + \frac{1}{M_\varepsilon}\rho_0\mu$$

and introducing this in (14), we find

$$\sigma_{ij} = \sigma^0_{ij} + (C^u_{ijkl} - \frac{1}{M_\varepsilon} a_{ij} a_{kl})\varepsilon_{kl} - \frac{1}{M_\varepsilon} a_{ij}\rho_0\mu.$$

A comparison with (58),(59) or (44),(62) then shows that the coefficients in these descriptions are related to those of the undrained description (14),(15) by

$$C_{ijkl} = C^u_{ijkl} - \frac{1}{M_\varepsilon} a_{ij} a_{kl}, \quad \alpha_{ij} = \frac{1}{M_\varepsilon} a_{ij}, \quad S_\varepsilon = \tilde{S}_\varepsilon + v_0 c_f = \frac{1}{M_\varepsilon}. \tag{63}$$

In particular, the relationship between the drained and the undrained stiffnesses may be expressed in the following ways

$$C^u_{ijkl} - C_{ijkl} = \alpha_{ij} a_{kl} = M_\varepsilon \alpha_{ij} \alpha_{kl} = S_\varepsilon a_{ij} a_{kl}. \tag{64}$$

The same relationship between C^u_{ijkl} and C_{ijkl}, expressed in terms of M_ε and α_{ij} was recorded earlier by Rudnicki (1985). It is of course an anticipated result, bearing in mind the formally analogous relations that exist in thermoelasticity between isothermal and adiabatic moduli[7].

Note also that the relations (63) will reduce to (50), as they must, if the fluid compressibility vanishes.

THE POTENTIAL $\chi(\sigma_{kl}, \mu)$. Finally, a further alternative drained description, based on the potential $\chi(\sigma_{kl}, \mu) = \varphi(\sigma_{kl}, m) - \mu m$, is obtained by rewriting the linear relationships (22),(23) in terms of the independent varible μ as

$$\varepsilon_{ij} = S_{ijkl}(\sigma_{kl} - \sigma^0_{kl}) + \beta_{ij} \rho_0 \mu, \tag{65}$$

and

$$\Delta m / \rho_0 = \beta_{ij}(\sigma_{ij} - \sigma^0_{ij}) + S_\sigma \rho_0 \mu, \tag{66}$$

observing that the compliance tensor must be the same as in (51), and that

$$\left.\frac{\partial \varepsilon_{ij}}{\partial p}\right|^0_{\sigma_{kl}} = \left.\frac{\partial \varepsilon_{ij}}{\partial \mu} \frac{\mathrm{d}\mu}{\mathrm{d}p}\right|^0_{\sigma_{kl}} = \left.\frac{1}{\rho_0} \frac{\partial \varepsilon_{ij}}{\partial \mu}\right|^0_{\sigma_{kl}} = \beta_{ij}$$

on account of (54). Assuming, as before, that the change in fluid mass content per unit reference volume can be approximated by (60), one has

$$\left.\frac{1}{\rho_0} \frac{\partial m}{\partial p}\right|^0_{\sigma_{kl}} = \left.\frac{1}{\rho_0} \frac{\partial m}{\partial \mu} \frac{\mathrm{d}\mu}{\mathrm{d}p}\right|^0_{\sigma_{kl}} = \left.\frac{\partial v}{\partial p}\right|^0_{\sigma_{kl}} + v_0 \left.\frac{1}{\rho_0} \frac{\partial \rho}{\partial p}\right|_0,$$

that is

$$S_\sigma = \tilde{S}_\sigma + v_0 c_f, \tag{67}$$

from which it is apparent that S_σ represents a *specific storage capacity at constant stress.* This interpretation of S_σ also clarifies the nature of

[7]Excellent discussions may be found in Nye (1957) and in Weiner (1983).

its reciprocal M_σ. We note further that S_σ is identical with the "storage compressibility" S of Kümpel (1991).

Again, we find that the constitutive relations (65),(66) may be cast alternatively in terms of the pore pressure change, yielding (51) and

$$\Delta m/\rho_0 = \beta_{ij}(\sigma_{ij} - \sigma_{ij}^0) + S_\sigma \Delta p. \tag{68}$$

as an alternative set.

We now put (23) into the form

$$\Delta m/\rho_0 = \frac{1}{M_\sigma} b_{kl}(\sigma_{kl} - \sigma_{kl}^0) + \frac{1}{M_\sigma} \rho_0 \mu$$

and introduce this in (22) to obtain

$$\varepsilon_{ij} = (S_{ijkl}^u - \frac{1}{M_\sigma} b_{ij} b_{kl})(\sigma_{kl} - \sigma_{kl}^0) - \frac{1}{M_\sigma} b_{ij} \rho_0 \mu.$$

The coefficients of the drained descriptions (65),(66) or (51),(68) and the undrained description (22),(23) are therefore related by

$$S_{ijkl} = S_{ijkl}^u - \frac{1}{M_\sigma} b_{ij} b_{kl}, \quad \beta_{ij} = \frac{1}{M_\sigma} b_{ij}, \quad S_\sigma = \tilde{S}_\sigma + v_0 c_f = \frac{1}{M_\sigma}. \tag{69}$$

In particular, the relationship between the drained und the undrained compliances may be expressed in the following ways

$$S_{ijkl} - S_{ijkl}^u = \beta_{ij} b_{kl} = M_\sigma \beta_{ij} \beta_{kl} = S_\sigma b_{ij} b_{kl}. \tag{70}$$

Note that the relations (69) reduce to (57), as they must, if the fluid compressibility vanishes.

A further observation, concerning equation (68), is that the quantity

$$\sigma_{ij} - \sigma_{ij}^0 + S_\sigma \beta_{ij}^{-1} \Delta p \tag{71}$$

plays the role of an *effective stress* governing the change in fluid mass content (here β_{ij}^{-1} denotes the inverse of β_{ij}). For isotropic materials, $S_\sigma/\beta_{ij}^{-1} = b_{ij}^{-1} = b^{-1}\delta_{ij} = (3/B)\delta_{ij}$, so that in this case (68) takes the form

$$\Delta m/\rho_0 = \beta \left(\sigma_{kk} - \sigma_{kk}^0 + \frac{3}{B} \Delta p \right) \tag{72}$$

in terms of Skempton's coefficient B, which is thus seen to govern the pore pressure response to changes in mean stress under undrained conditions

$$\Delta p = -B \frac{\sigma_{kk} - \sigma_{kk}^0}{3}. \tag{73}$$

As was shown by Rice and Cleary (1976), the fluid mass content (72) of a slightly compressible fluid satisfies the homogeneous diffusion equation[8]

$$c\nabla^2 m = \partial m/\partial t, \tag{74}$$

in which c is a *coefficient of consolidation* or *hydraulic diffusivity* that is given by (cf. Eq. 17 in Rice and Cleary (1976))

$$c = \frac{k}{\eta}\left[\frac{2GB^2(1+\nu_u)^2(1-\nu)}{9(1-\nu_u)(\nu_u-\nu)}\right], \tag{75}$$

and where k is the absolute permeability[9] of the (isotropic) medium (in m^2) and η is the viscosity of its pore fluid (in Pa s). Also, ν and ν_u denote the drained and undrained Poisson ratio, respectively, as defined and discussed further below. Making use of the relationships given there for isotropic materials and the above definitions of the storage parameters S_σ and S_ε, it is easily seen that the hydraulic diffusivity can be defined alternatively by the following expressions in terms of the parameters α, β, ν and S_σ or S_ε (see Eqs. 112–117 further below)

$$c = \frac{k/\eta}{S_\sigma - [2(1-2\nu)/(1-\nu)]\alpha\beta} = \frac{k/\eta}{S_\varepsilon + [(1+\nu)/(1-\nu)]\alpha\beta}. \tag{76}$$

The quantity in brackets in (75) thus represents the reciprocal of

$$S = S_\sigma - [2(1-2\nu)/(1-\nu)]\alpha\beta = S_\varepsilon + [(1+\nu)/(1-\nu)]\alpha\beta, \tag{77}$$

which provides two of several equivalent expressions for a *specific storage capacity* or *storage coefficient*[10]. Note, however, that elsewhere in the literature the symbol S is often used to denote a different storage parameter (see, e.g., Wang (2000). Also, Green and Wang (1990) define a *specific storage coefficient* $S_s = \rho_f g S$ as the product of (77) and the unit weight of water.

[8] Note that (74) becomes a rigorous result for incompressible pore fluids, when in effect it reduces to a diffusion equation for the change Δv in pore volume fraction.

[9] In order to avoid any confusion with the elastic bulk modulus K, we denote the absolute permeability by the lower-case symbol k; we thereby depart from the standard notation K in the hydrological literature, where the lower-case letter is usually reserved for the so-called *coefficient of permeability* $k = \rho g K/\eta$ with the dimension of a velocity; ρ and η then denote the uniformly constant density and viscosity of the groundwater.

[10] See also Jaeger *et al.* (2007, chap. 7) on this point.

Relationships between drained coefficients. From (46) and (53) it follows by an application of the chain rule that

$$S_{ijkl}\,C_{klmn} = \frac{\partial \varepsilon_{ij}}{\partial \sigma_{kl}}\bigg|_p^0 \frac{\partial \sigma_{kl}}{\partial \varepsilon_{mn}}\bigg|_p^0 = \frac{1}{2}\left(\delta_{im}\delta_{jn} + \delta_{in}\delta_{jm}\right). \tag{78}$$

Substitution of this result in (44) also yields

$$S_{ijkl}\,C_{klmn}\,\varepsilon_{mn} = \varepsilon_{ij} = S_{ijkl}(\sigma_{kl} - \sigma^0_{ijkl}) + S_{ijkl}\,\alpha_{kl}\Delta p$$

and, upon comparing this with (51),

$$\beta_{ij} = S_{ijkl}\,\alpha_{kl}. \tag{79}$$

In particular, if the porous medium is isotropic, its response is characterized by the isotropic tensors $\alpha_{ij} = \alpha\delta_{ij}$, $\beta_{ij} = \beta\delta_{ij}$ and $S_{ijkl} = \frac{1}{4G}(\delta_{ik}\delta_{jl} + \delta_{il}\delta_{jk}) + (\frac{1}{9K} - \frac{1}{6G})\delta_{ij}\delta_{kl}$, where K is the *drained* bulk modulus and G is the shear modulus of the of the porous solid. The last relation therefore reduces to (cf. Biot (1941), Eq. 2.11)

$$\beta = \frac{1}{3}S_{iikk}\,\alpha = \frac{\alpha}{3K}. \tag{80}$$

Similarly, one finds that

$$C_{ijkl}\,S_{klmn} = \frac{1}{2}\left(\delta_{im}\delta_{jn} + \delta_{in}\delta_{jm}\right) \tag{81}$$

Using this property in (51), after multiplication by C_{ijkl}, gives

$$\sigma_{ij} - \sigma^0_{ij} = C_{ijkl}\,\varepsilon_{kl} - C_{ijkl}\,\beta_{kl}\Delta p$$

and, upon comparing this with (44),

$$\alpha_{ij} = C_{ijkl}\,\beta_{kl}. \tag{82}$$

In the case of isotropy, with $C_{ijkl} = G(\delta_{ik}\delta_{jl} + \delta_{il}\delta_{jk}) + (K - 2G/3)\delta_{ij}\delta_{kl}$ and $C_{iikk} = 9K$, this again reduces to (80).

From (44),(45), and (82), it also follows that

$$v - v_0 = C_{ijkl}\,\beta_{kl}\,\varepsilon_{ij} + \tilde{S}_\varepsilon\Delta p = \beta_{ij}(\sigma_{ij} - \sigma^0_{ij}) + (\alpha_{ij}\beta_{ij} + \tilde{S}_\varepsilon)\Delta p$$

and hence, from (52),(61),(67),(79), and (82),

$$\tilde{S}_\sigma - \tilde{S}_\varepsilon = S_\sigma - S_\varepsilon = \alpha_{ij}\,\beta_{ij} = \alpha_{ij}\,S_{ijkl}\,\alpha_{kl} = \beta_{ij}\,C_{ijkl}\,\beta_{kl}. \tag{83}$$

For an isotropic porous medium one obtains (cf. Biot (1941), Eq. 2.12)

$$\tilde{S}_\sigma - \tilde{S}_\varepsilon = S_\sigma - S_\varepsilon = 3\alpha\beta = K^{-1}\alpha^2 = 9K\beta^2. \tag{84}$$

3.3 Diagrammatic Representation of Constitutive Relations; Summary of Relationships Between Constitutive Coefficients

The entire set of linear constitutive relations derived in the above can be displayed in the form of a diagram by positioning the two tensorial field variables $\Delta\sigma_{ij}$ and ε_{ij} and the two pairs of (alternative) scalar field variables $\Delta p, \rho_0\mu$ and $\Delta v, \Delta m/\rho_0$ at diagonally opposite corners of a square (Fig. 1). Every pair of constitutive relations comprises one tensor- and one scalar-valued equation in one tensorial and one scalar independent variable, as for example the set formed by (44) & (45), which we may write

$$\begin{aligned} \Delta\boldsymbol{\sigma} &= \mathbb{C}:\boldsymbol{\varepsilon} - \boldsymbol{\alpha}\,\Delta p, \\ \Delta v &= \boldsymbol{\alpha}:\boldsymbol{\varepsilon} + \tilde{S}_\varepsilon \Delta p, \end{aligned}$$

in direct tensor notation[11]. These equations are extracted from Figure 1 by selecting the pair of dependent variables $\Delta\boldsymbol{\sigma}, \Delta v$ and the pair of independent variables $\boldsymbol{\varepsilon}, \Delta p$ and by factoring the latter with the coefficients $\mathbb{C}, \boldsymbol{\alpha}, \tilde{S}_\varepsilon$ that label the arrows pointing *from the independent to the dependent variable* along a connecting line.

Here the following rules must be observed: A subscript ϵ or σ must be attached to the scalar coefficient S and its reciprocal $M = 1/S$ in all cases as an indication of the chosen independent tensorial variable. If Δp or $\rho_0\mu$ are selected as dependent variables, the rank-4 tensors $\mathbb{C}$ or $\mathbb{S}$ are to be viewed as *drained coefficients* and these carry no superscript. On the other hand, if $\Delta m/\rho_0$ or Δv are selected as dependent variables, then the *undrained coefficients* $\mathbb{C}^u$ or $\mathbb{S}^u$ apply. Accordingly, for example, one recovers (22) & (23)

$$\begin{aligned} \boldsymbol{\varepsilon} &= \mathbb{S}^u:\Delta\boldsymbol{\sigma} + \mathbf{b}\,\Delta m/\rho_0, \\ \rho_0\mu &= -\mathbf{b}:\Delta\boldsymbol{\sigma} + M_\sigma \Delta m/\rho_0, \end{aligned}$$

upon selecting $\Delta\boldsymbol{\sigma}$ and $\Delta m/\rho_0$ as independent variables. However, choosing Δv as independent variable will require a tilde to be placed upon all coefficients for such special undrained descriptions, yielding for example (27) & (28)

$$\begin{aligned} \Delta\boldsymbol{\sigma} &= \tilde{\mathbb{C}}^u:\boldsymbol{\varepsilon} - \tilde{\mathbf{a}}\,\Delta v, \\ \Delta p &= -\tilde{\mathbf{a}}:\boldsymbol{\varepsilon} + \tilde{M}_\varepsilon \Delta v. \end{aligned}$$

[11] Here the double-dot product $\boldsymbol{\alpha}:\boldsymbol{\varepsilon} = \alpha_{ij}\varepsilon_{ij}$ denotes the scalar product of two rank-2 tensors, $\mathbb{C}:\boldsymbol{\varepsilon}$ the double contraction of rank-4 tensor $\mathbb{C}$ with rank-2 tensor $\boldsymbol{\varepsilon}$ giving the rank-2 tensor with components $C_{ijkl}\varepsilon_{kl}$.

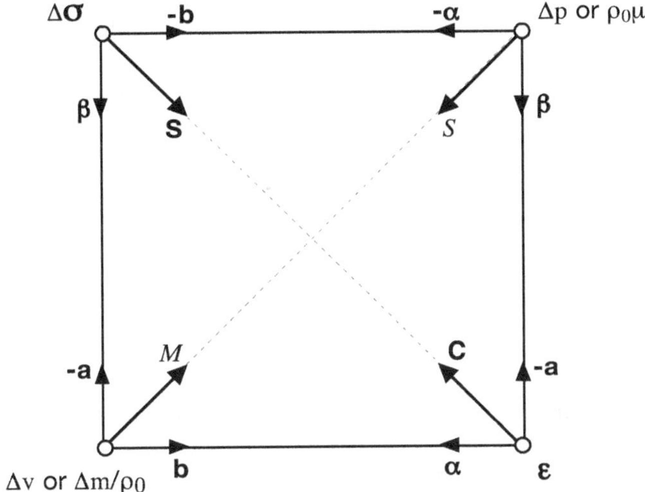

Figure 1. Diagrammatic representation of constitutive relations.

We note further that the above constitutive relation (45) will be recovered from the last equation upon switching the role of the variables Δp and Δv. This particular change of dependent and independent variable transforms, as we have seen, the last undrained description into the preceding drained description via the relationships (50), i.e., through $\tilde{\mathbb{C}}^u - \mathbb{C} = \boldsymbol{\alpha}\tilde{\mathbf{a}}$, $\tilde{\mathbf{a}} = \tilde{M}_\varepsilon \boldsymbol{\alpha}$, and $\tilde{M}_\varepsilon = 1/\tilde{S}_\varepsilon$ between the respective coefficients. The appearance of $\tilde{S}_\varepsilon$ or $\tilde{S}_\sigma$ in a drained description is thus explained by the fact that these coefficients relate the pore volume change to a pressure change at constant strain or stress and thus capture the response of the 'wetted' porous skeleton alone.

Table 2 provides a summary of the above-derived relationships between the constitutive coefficients of alternative descriptions. The relationships between the drained and undrained compliances or stiffnesses of a poroelastic material are of particular interest, the first result of this kind having been obtained by Gassmann (1951). Because of the practical interest that attaches to Gassmann's relations, we shall examine them in more detail in the remainder of this review, paying attention to various alternative expressions in terms of particular set of material parameters that can be obtained from so-called *unjacketed tests.*

The pair-structure of the constitutive theory of poroelasticity, comprising one tensorial and one scalar relation and *three* constitutive coefficients of different tensorial rank, is not always brought out clearly in the literature. Thompson and Willis (1991), for example, derive a relationship for $\Delta m/\rho_0$ (their equation 32), in which the coefficient of the pore fluid pressure becomes $S_{jjkl}\alpha_{kl}/b_{ii} - \alpha_{ij}S_{ijkl}\alpha_{kl}$. But this is simply an expression for the scalar coefficient S_ε, as is easily seen from the relationships listed in Table 2, so that their equation 32 is the same as (62); its existence could thus have been inferred directly from the relevant thermodynamic potential $\pi(\varepsilon_{kl}, \mu)$.

Table 2 and Figure 1 provide readers with a summary of the information needed to gain a rapid overview over the various possible alternative formulations of linear poroelasticity. As has been emphasized in the above at various places, the fundamental choice that one faces in practice is between a formulation in terms of the properties of Biot's "wetted solid" and a formulation that will include properties of the pore fluid, viz. a constant fluid compressibility, in the constitutive description. Looking again at Table 1 one would, after all that has been said in the above, most likely come up with a choice of either a drained description based on the potentials $\tilde{\pi}(\varepsilon_{kl}, p)$ or $\tilde{\chi}(\sigma_{kl}, p)$ or an undrained description based on the potentials $\psi(\varepsilon_{kl}, m)$ or $\phi(\sigma_{kl}, m)$. As a linear theory, the latter would demand a constant fluid compressibility, as we have seen. If one wished to preserve the linear elastic response of a wetted porous solid, but at the same time allow for the presence of a highly compressible fluid phase and large changes in fluid pressure, then a drained description will be the appropriate choice, since it allows us to defer any account of fluid behaviour to the formulation of the final field equations of the theory. An undrained description in terms the potential $\tilde{\psi}(\varepsilon_{kl}, v)$ would offer the same advantage, were it not for the impractical choice of the volume fraction as independent variable. The potential $\tilde{\psi}(\varepsilon_{kl}, v)$ nevertheless furnishes the appropriate point of departure in constructing a linear theory, as we have seen. It was selected by Biot (1941) for this purpose. Biot however rewrote the resulting constitutive relations in terms of the stress and the pore pressure as independent variables, which amounts to writing the (isotropic) equivalent of our Eqs. (51),(52) rather than (34),(35).

Table 2. Relations between constitutive coefficients of linear poroelasticity

Relation	Equation number
$S_{ijkl}C_{klmn} = \frac{1}{2}(\delta_{im}\delta_{jn} + \delta_{in}\delta_{jm})$	(78)
$\beta_{ij} = S_{ijkl}\alpha_{kl}, \ \alpha_{ij} = C_{ijkl}\beta_{kl}$	(79,82)
$S_\sigma - S_\varepsilon = \tilde{S}_\sigma - \tilde{S}_\varepsilon = \alpha_{ij}\beta_{ij} = \alpha_{ij}S_{ijkl}\alpha_{kl} = \beta_{ij}C_{ijkl}\beta_{kl}$	(83)
$S_\varepsilon = M_\varepsilon^{-1} = \tilde{S}_\varepsilon + v_0 c_f, \ S_\sigma = M_\sigma^{-1} = \tilde{S}_\sigma + v_0 c_f$	(61,67)
$S^u_{ijkl}C^u_{klmn} = \frac{1}{2}(\delta_{im}\delta_{jn} + \delta_{in}\delta_{jm})$	(39)
$b_{ij} = S^u_{ijkl}a_{kl}, \ a_{ij} = C^u_{ijkl}b_{kl}$	(40,42)
$M_\varepsilon - M_\sigma = a_{ij}b_{ij} = a_{ij}S^u_{ijkl}a_{kl} = b_{ij}C^u_{ijkl}b_{kl}$	(43)
$a_{ij} = M_\varepsilon\alpha_{ij}, \ b_{ij} = M_\sigma\beta_{ij}$	(63,69)
$C^u_{ijkl} - C_{ijkl} = \alpha_{ij}a_{kl} = M_\varepsilon\alpha_{ij}\alpha_{kl} = S_\varepsilon a_{ij}a_{kl}$	(64)
$S_{ijkl} - S^u_{ijkl} = \beta_{ij}b_{kl} = M_\sigma\beta_{ij}\beta_{kl} = S_\sigma b_{ij}b_{kl}$	(70)

4 Elastic Bulk Moduli from Unjacketed Tests

The sets of coefficients $C_{ijkl}, \alpha_{ij}, \tilde{S}_\sigma$ and $S_{ijkl}, \beta_{ij}, \tilde{S}_\varepsilon$ which have been introduced along with the pairs of state variables ε_{ij}, p and σ_{ij}, p, respectively, provide alternative characterizations of linear poroelastic behaviour. Certain of these parameters may however be difficult to determine experimentally. It is therefore important to consider other, possibly more accessible parameters as primary parameters that are defined by certain combinations of the above coefficients, these combinations being suggested by a particular type of experiment.

Here we shall consider one such possibility that arises naturally with the idea of a so-called unjacketed compression test, which has occupied a prominent position in discussions of poroelastic constants(Geertsma, 1957b; Nur and Byerlee, 1971; Biot, 1973; Brown and Korringa, 1975; Rice and Cleary, 1976; Carroll, 1979; Zimmerman *et al.*, 1986; Kümpel, 1991; Detournay and Cheng, 1993; Berge and Berryman, 1995; Coussy, 2004; Wang, 2000). In an unjacketed test, a macroscopically homogeneous, saturated sample is subjected to a change Δp in pore fluid pressure and simultaneous change $\Delta\sigma_{ij} \equiv \sigma_{ij} - \sigma_{ij}^0 = -\Delta p\,\delta_{ij}$ of total stress on its faces[12]. Such an experiment will yield changes $V - V_0$ and $V_\phi - V_{\phi 0}$ in the bulk and pore volume of the sample that are linked through the relation $V - V_0 = V_\phi - V_{\phi 0} + V_s - V_{s0}$ to the change $V_s - V_{s0}$ in solid skeleton volume. One may therefore distinguish three different compressibilities or bulk moduli[13] for such an unjacketed test which, in the linear approximation, must be related by (Brown and Korringa, 1975)

$$\frac{1}{K_s'} = \frac{v_0}{K_s''} + \frac{1 - v_0}{K_s}, \tag{85}$$

where

[12] This type of hydrostatic loading of a porous solid is also referred to as Π-*loading* in the literature on the subject (Detournay and Cheng, 1993).

[13] We use the notation introduced by Rice and Cleary (1976) for these moduli. Their modulus K_s'' is identical with the modulus K_ϕ as defined by Brown and Korringa (1975), Berge and Berryman (1995), Wang (2000), and others. Mavko *et al.* (1998), on the other hand, define K_ϕ differently as *dry pore space stiffness*, the reciprocal of which corresponds to $3\beta = 1/K - 1/K_s'$ in our notation; the same authors usually assume microscopic homogeneity and equate all three moduli in (86)–(88) to a mineral bulk modulus K_s which they chose to denote by K_0. Guéguen *et al.* (2004) also make this assumption. Detournay and Cheng (1993) use yet another definition, putting $1/K_\phi = 1/K - 1/(K_s' v_s)$, where v_s is the volume fraction of the solid phase.

$$\frac{1}{K_s'} = -\frac{1}{V_0}\frac{\partial V}{\partial p}\bigg|_{\Delta\sigma_{ij}=-\Delta p\delta_{ij}} = -\frac{\varepsilon_{kk}}{\Delta p}, \tag{86}$$

$$\frac{1}{K_s''} = -\frac{1}{V_{\phi 0}}\frac{\partial V_\phi}{\partial p}\bigg|_{\Delta\sigma_{ij}=-\Delta p\delta_{ij}} = -\frac{1}{v_0}\frac{v - v_0}{\Delta p}, \tag{87}$$

$$\frac{1}{K_s} = -\frac{1}{V_{s0}}\frac{\partial V_s}{\partial p}\bigg|_{\Delta\sigma_{ij}=-\Delta p\delta_{ij}} = -\frac{1}{1-v_0}\frac{v_s - v_{s0}}{\Delta p}. \tag{88}$$

The fact that these moduli are defined as properties of a representative macroscopic rock sample implies in particular that K_s is to be viewed as an average bulk modulus of a skeleton body which in general may be quite heterogeneous and may also contain fluid inclusions.

The relationships between the moduli K_s, K_s', K_s'' and the earlier defined poroelastic constants are now readily established by introducing the stress change $\Delta\sigma_{ij} = -\Delta p\,\delta_{ij}$, appropriate for the unjacketed test, in relation (51) and (52). Making use of (86) and (87) this gives

$$\beta_{ii} = \frac{1}{K} - \frac{1}{K_s'} \tag{89}$$

and

$$\tilde{S}_\sigma = \beta_{ii} - \frac{v_0}{K_s''}, \tag{90}$$

and consequently

$$\tilde{S}_\sigma = \frac{1}{K} - \frac{1}{K_s'} - \frac{v_0}{K_s''}. \tag{91}$$

Here we have written $S_{iikk} = 1/K$ for the drained bulk volume compressibility. Although the notation K is usually reserved for the bulk modulus of elastically isotropic materials, its use for anisotropic materials is justified by the invariance of the sum S_{iikk} under changes of coordinate axes.

The interest in the above relationships stems from the possibility to determine the coefficients in the linear constitutive relations (44),(45) or (51),(52) at least partly in terms of properties obtained from unjacketed tests. It would of course be preferable to measure the scalar parameter $\tilde{S}_\sigma$ directly, but the practical difficulty of such a measurement suggests the use of expression (91) in terms of other parameters.

In the special situation of a strictly homogeneous and isotropic solid skeleton Π-loading will induce a uniform pressure Δp throughout the solid phase, which will therefore undergo a spatially uniform volume change equal to the relative change in bulk volume and the relative change in pore volume, this change being quite independent of pore shape and volume fraction. It follows that $K_s' = K_s'' = K_s$ in this case, where K_s is the uniform bulk modulus of the solid phase (Geertsma, 1957b; Nur and Byerlee, 1971). To reach this conclusion, it suffices to add the term $-(1-v_0)/K_s'$ to both sides of equation 85 and thereafter to use the defining relations (86)–(88) to write

$$\begin{aligned} v_0\left(\frac{1}{K_s'}-\frac{1}{K_s''}\right) &= (1-v_0)\left(\frac{1}{K_s}-\frac{1}{K_s'}\right) \\ &= \left(\frac{V_\phi}{V_{\phi 0}}-\frac{V}{V_0}\right)\left(\frac{V_{\phi 0}}{V_0}\right)(\Delta p)^{-1} = \left(\frac{V}{V_0}-\frac{V_s}{V_{s0}}\right)\left(\frac{V_{s0}}{V_0}\right)(\Delta p)^{-1}. \end{aligned}$$

We conclude that $K_s' = K_s'' = K_s$ whenever $V/V_0 = V_\phi/V_{\phi 0} = V_s/V_{s0}$. That the latter conditions must be satisfied for Π-loading of a microscopically homogeneous and isotropic porous solid (with interconnected pores only) follows from a *Gedankenexperiment* in which one envisages at first a hydrostatically loaded, uniformly compressed non-porous solid from which the pore-space is subsequently cut out (Gassmann, 1951; Nur and Byerlee, 1971; Brown and Korringa, 1975). If the uniform pressure along these imagined cuts is restored after the cutting operation by the application of a pore fluid pressure of equal magnitude, then the initially homogeneous state of strain in the solid phase will remain the same before and after the cutting, as will the shape of the pore space. Π-loading therefore leads to the same relative change in all linear dimensions of the skeleton, the pore space, and the bulk of the porous continuum[14].

Carroll (1979) has subsequently generalized this argument, assuming the solid skeleton to be microscopically homogeneous but anisotropic. Carroll's analysis leads to the following result (in our notation) for the pore pressure coefficient α_{ij} in (14) and (49):

$$\alpha_{ij} = \delta_{ij} - C_{ijkl} S^s_{klmm} \tag{92}$$

where S^s_{klmn} denotes the tensor of elastic compliances of the microscopically homogeneous solid material.

It is clear, however, that the assumption of microscopic homogeneity will be rather restrictive and artificial in most cases. Thompson and Willis (1991) have therefore proposed a reformulation of the theory in which the pore pressure coefficient β_{ij} is expressed in terms of some *average* elastic compliance S'_{ijkl} of the solid skeleton material, which is *defined* to satisfy the relation

$$S'_{ijkk} = S_{ijkk} - \beta_{ij}, \tag{93}$$

[14] An unjacketed compression test on an elastically homogeneous porous skeleton therefore produces no change in porosity.

so that on account of (82) α_{ij} is given by

$$\alpha_{ij} = \delta_{ij} - C_{ijkl}S'_{klmm}. \tag{94}$$

This is of the same form as (92), but the microscopically uniform compliance S^s_{ijkl} is now replaced by an effective average solid matrix compliance S'_{ijkl}. Note that $S'_{iikk} = 1/K'_s$, by virtue of (89), and this compressibility differs in general from $S^s_{iikk} = 1/K_s$.

Each of the moduli K'_s, K''_s, and K_s has been defined operationally in the above by a certain macroscopic experiment. In the general case of a microscopically heterogeneous and macroscopically anisotropic skeleton, these moduli are therefore still well-defined by (86)–(88), but remain related, in the first place, through (85) only. Micro-mechanical considerations enter, for example, with the homogeneity requirement in the above demonstration of the equalities $K'_s = K''_s = K_s$ (Nur and Byerlee, 1971). If a porous skeleton is composed of one very stiff solid component and a second very compressible component, then a uniform increase in the external and internal (pore fluid) pressure may lead to an increase in the pore volume at the expense of the volume of the very compressible component, in which case K''_s will be negative (Berge and Berryman, 1995) (see also Zimmerman *et al.* (1986) and Berryman (1995) for discussions of bounds on the magnitudes of various poroelastic constants).

5 Isotropic Response

5.1 Isotropic Unjacketed Response

The significance of the compliance S'_{ijkl} is further clarified by considering the special case of Π-*loading*, when $\Delta\sigma_{ij} = -\Delta p\delta_{ij}$ and equation 51 yields

$$\varepsilon_{ij} = -(S_{ijkk} - \beta_{ij})\Delta p = -S'_{ijkk}\Delta p. \tag{95}$$

From this it is apparent that Π-loading constitutes an experimental strategy by which the average anisotropy of the skeleton material, as embodied in the compliance S'_{ijkl}, may be separated and isolated from the textural (microstructure-induced) anisotropy of the rock, although it is the component sums S'_{ijkk} rather than the full tensor S'_{ijkl} that will be obtained in this way.

The special case of an isotropic unjacketed response merits some attention at this point because of its potential practical importance. This type of response to Π-loading produces the isotropic bulk strain

$$\varepsilon_{ij} = \frac{1}{3}\varepsilon_{kk}\,\delta_{ij} = -(S_{ijkk} - \beta_{ij})\Delta p = -S'_{ijkk}\Delta p. \tag{96}$$

From (86) it then follows that

$$S'_{ijkk} = \frac{1}{3K'_s}\delta_{ij}. \tag{97}$$

In the case of an isotropic unjacketed response, one therefore has the relationship

$$\beta_{ij} = S_{ijkk} - \frac{1}{3K'_s}\delta_{ij}, \tag{98}$$

and, after substitution in (82) and use of (81),

$$\alpha_{ij} = \delta_{ij} - \frac{1}{3K'_s}C_{ijkk}. \tag{99}$$

Expressions of the same form for the pore pressure coefficients could of course have been obtained directly from Carroll's result by putting $S^s_{klmm} = (1/K_s)\delta_{kl}$ in (92). However, the modulus K'_s would then have to be equated to the bulk modulus K_s of a homogeneous and isotropic skeleton material, while its interpretation in (98) and (99) is that of a bulk modulus of a possibly heterogeneous and locally anisotropic matrix that is sufficiently disordered to display, on average, an isotropic strain response to Π-loading.

We note further that the off-diagonal components of β_{ij} are entirely determined by the component sums S_{ijkk} of the compliance, while the trace β_{ii} assumes the same form (89) as in the general case of anisotropy. Also, according to (99), the components α_{ij} are all expressible in terms of a component sum C_{ijkk} multiplied by the same scalar factor $1/3K'_s$.

An isotropic unjacketed response to Π–loading implies that (45) becomes

$$\Delta v = \alpha_{ii}\frac{1}{3}\varepsilon_{kk} + \tilde{S}_\varepsilon \Delta p. \tag{100}$$

Dividing this by Δp, making use of definitions (86),(87) and substituting the appropriate expression for α_{ii} from (99), we obtain

$$\tilde{S}_\varepsilon = \frac{1}{K'_s}\left(1 - \frac{K}{K'_s}\right) - \frac{v_0}{K''_s}, \tag{101}$$

having put $C_{iikk} = 9K$.

5.2 Fully Isotropic Macroscopic Response

For the isotropic forms of the rank-2 and rank-4 tensor coefficients, which we recorded in deriving (80) and (84), relations (64) and (70) reduce to

$$(G_u - G)(\delta_{ik}\delta_{jl} + \delta_{il}\delta_{jk} - \frac{2}{3}\delta_{ij}\delta_{kl}) + (K_u - K)\delta_{ij}\delta_{kl} = \alpha a\, \delta_{ij}\delta_{kl} \tag{102}$$

and

$$\left(\frac{1}{4G}-\frac{1}{4G_u}\right)(\delta_{ik}\delta_{jl}+\delta_{il}\delta_{jk}-\frac{2}{3}\delta_{ij}\delta_{kl})+\left(\frac{1}{9K}-\frac{1}{9K_u}\right)\delta_{ij}\delta_{kl}=\beta b\,\delta_{ij}\delta_{kl}, \tag{103}$$

respectively, and these differences vanish, whenever we have $i \neq j$, $k \neq l$ or both. In particular, for any fixed pair of subscripts r, s, such that $i = k = r$, $j = l = s$ and $r \neq s$, we obtain

$$C^u_{rsrs} - C_{rsrs} = G_u - G = 0 \text{ (not summed over repeated subscripts)} \tag{104}$$

Thus the elastic shear modulus is the same for drained and undrained deformations. On the other hand, letting $i = j$ and $k = l$ in (102) and (103), it is found that

$$K_u - K = \alpha a = \frac{\alpha^2}{S_\varepsilon} \tag{105}$$

and

$$\frac{1}{K} - \frac{1}{K_u} = 9\beta b = \frac{9\beta^2}{S_\sigma}, \tag{106}$$

where we have made use of (63) and (69). Moreover, dividing (106) by (105) and using (80), we obtain

$$\frac{K_u}{K} = \frac{S_\sigma}{S_\varepsilon} \tag{107}$$

as a further useful relationship and counterpart of the well-known thermodynamic relation $c_p/c_v = \kappa_T/\kappa_S$ between the ratio of the specific heats at constant pressure and volume and the ratio of the isothermal and adiabatic compressibilities.

Corresponding results for the differences between the Poisson ratios ν and ν_u and Young's moduli E and E_u may now be obtained by expressing K (K_u) in terms of G and ν (ν_u) in (106)[15]. This gives

$$\nu_u = \frac{3\nu + (1-2\nu)\alpha B}{3 - (1-2\nu)\alpha B} \tag{108}$$

in agreement with the result of Rice & Cleary (1976). For Young's modulus, one uses the relationship $E = 2(1+\nu)G$, together with (104), to deduce

$$\frac{E_u}{E} = \frac{1+\nu_u}{1+\nu} \tag{109}$$

[15]The expression used is $3K = 2G(1+\nu)/(1-2\nu)$.

Further, from (89) and (90) it is seen that in the case of isotropy

$$\beta = \frac{1}{3}\left(\frac{1}{K} - \frac{1}{K_s'}\right) \tag{110}$$

and hence

$$\tilde{S}_\sigma = 3\beta - \frac{v_0}{K_s''} = \frac{1}{K} - \frac{1}{K_s'} - \frac{v_0}{K_s''} \tag{111}$$

must hold; consequently one has

$$\frac{1}{M_\sigma} = S_\sigma = \tilde{S}_\sigma + v_0 c_f = \frac{1}{K} - \frac{1}{K_s'} + v_0\left(\frac{1}{K_f} - \frac{1}{K_s''}\right), \tag{112}$$

where we have introduced the notation K_f for the bulk modulus c_f^{-1} of the fluid phase. For fully isotropic materials $(69)_2$, (110), and (112) thus lead to the following expressions for the Skempton coefficient (Rice and Cleary, 1976)

$$B = 3b = 3M_\sigma\beta = \frac{3\beta}{\tilde{S}_\sigma + v_0 c_f} = \frac{1/K - 1/K_s'}{1/K - 1/K_s' + v_0(1/K_f - 1/K_s'')} \tag{113}$$

From (106) and (110) also follows the expression for B in terms of the bulk moduli

$$B = \frac{1/K - 1/K_u}{1/K - 1/K_s'} \tag{114}$$

which has been recorded by several authors, albeit with K_s' equated to K_s.

Substitution of (110) for β in (80) now yields the familiar isotropic form of the Biot coefficient

$$\alpha = 1 - \frac{K}{K_s'}. \tag{115}$$

Using this in (84) together with (111), one gets

$$\tilde{S}_\varepsilon = \frac{\alpha}{K_s'} - \frac{v_0}{K_s''} = \frac{1}{K_s'}\left(1 - \frac{K}{K_s'}\right) - \frac{v_0}{K_s''} \tag{116}$$

and therefore

$$\frac{1}{M_\varepsilon} = S_\varepsilon = \tilde{S}_\varepsilon + v_0 c_f = \frac{1}{K_s'}\left(1 - \frac{K}{K_s'}\right) + v_0\left(\frac{1}{K_f} - \frac{1}{K_s''}\right). \tag{117}$$

The scalar coefficient a can thus be expressed in the following forms

$$a = M_\varepsilon\alpha = \frac{\alpha}{\tilde{S}_\varepsilon + v_0 c_f} = \frac{1 - K/K_s'}{(1/K_s')(1 - K/K_s') + v_0(1/K_f - 1/K_s'')}. \tag{118}$$

5.3 Comparison of Parameter Definitions for Fully Isotropic Materials

At certain points in the above we have already identified the constitutive coefficients defined by Biot in his 1941 paper with corresponding parameters of our formulation. This comparison is carried somewhat further in Table 3 for the special case of isotropy, by including equivalent quantities as defined by Rice and Cleary (1976) whose formulation has frequently been compared with others; readers should be able to construct further equivalences using such sources together with the information provided by Table 3.

In his early work, Biot conceived the variable θ as *increment of water volume per unit volume of soil*, calling it "variation in water content". Biot's 1941 paper is essentially a linear theory for a wetted poroelastic skeleton, involving the variation in water content θ and the *increment in water pressure* σ as work-conjugate variables. Biot's θ should therefore be interpreted as the equivalent of the relative volume change $v - v_0$ in the formulation of Rice and Cleary (1976) as well as in the present one. This implies that Biot's parameter $1/R$ should be interpreted as the equivalent of our $\tilde{S}_\sigma$ rather than that of S_σ.[16] As shown in Table 3, this interpretation is consistent with that of Rice and Cleary, but differs from the "corrected" interpretation of Green and Wang (1986) and Wang (2000) that would equate Biot's $1/R$ to our S_σ and demand the addition of a "missing term" v_0/K_s' to the original expression in Rice and Cleary (1976). Here we suggest to reverse this "correction", i.e., to restore the original interpretation by Rice and Cleary and in this way to recover a drained description of linear poroelasticity that makes no reference to fluid properties. The need for such a description arises with the formulation of non-linear field equations for poroelastic materials that are saturated by highly compressible pore fluids. It is clear that Green and Wang's "correction" follows from their view of Biot's potential U as a function of the strain and fluid mass content m, rather than of the strain and the pore volume change, contrary to Biot's explicit statement (see, e.g., his Eq. 2.7).[17] Green and Wang's formulation remains of course a possible one and in the present context implies the choice of (68) rather than (52) as a scalar constitutive relation to go along with (51). In consequence, however, one must assume a constant fluid compressibility that will enter into expression (67) for S_σ.

[16]This agrees with the interpretation of Detournay and Cheng (1993), who use R', H', and Q' to denote R, H, and Q as defined by Biot (1941). Subsequently, however, Biot (1955) used the same symbols for different quantities.

[17]It is unfortunate that Wang (2000) also fails to distinguish clearly between constitutive relations involving the variable m from those involving the variable v.

Table 3. Equivalent parameter definitions in isotropic poroelasticity

Biot (1941)	Rice & Cleary (1976)	This chapter	(Eqn. No.)
-	$(m-m_0)/\rho_0$	$(m-m_0)/\rho_0$	(2)
θ	$v-v_0$	$v-v_0$	(4)
$1/Q$	$\frac{1}{K_s'}\left(1-\frac{K}{K_s'}\right)-\frac{v_0}{K_s''}$	$\tilde{S}_\varepsilon=\tilde{M_\varepsilon}^{-1}$	(31,48)
$1/R$	$\frac{1}{K}-\frac{1}{K_s'}-\frac{v_0}{K_s''}$	$\tilde{S}_\sigma=\tilde{M_\sigma}^{-1}$	(38,55)
-	$\frac{1}{K_s'}(1-\frac{K}{K_s'})-\frac{v_0}{K_s''}+\frac{v_0}{K_f}$	$S_\varepsilon=M_\varepsilon^{-1}$	(18,61)
-	$\frac{1}{K}-\frac{1}{K_s'}-\frac{v_0}{K_s''}+\frac{v_0}{K_f}$	$S_\sigma=M_\sigma^{-1}$	(26,67)
-	$\frac{1}{K_f}$	$\frac{1}{K_f}=c_f$	
-	$\frac{1}{K_s'}$	$\frac{1}{K_s'}$	(86)
-	$\frac{1}{K_s''}$	$\frac{1}{K_s''}$	(87)
-	$\frac{1}{K_s}$	$\frac{1}{K_s}$	(88)
α	$1-\frac{K}{K_s'}$	α	(47)
$1/H$	$\frac{1}{K}-\frac{1}{K_s'}$	3β	(54)
αQ	$\frac{1-K/K_s'}{(1-K/K_s')/K_s'-v_0/K_s''}$	$\tilde{a}$	(30)
R/H	$\frac{1/K-1/K_s'}{1/K-1/K_s'-v_0/K_s''}$	$3\tilde{b}$	(37)
-	$\frac{1-K/K_s'}{(1-K/K_s')/K_s'-v_0/K_s''+v_0/K_f}$	a	(17)
-	$\frac{1/K-/K_s'}{1/K-1/K_s'-v_0/K_s''+v_0/K_f}$ $(=B)$	$3b$ $(=B)$	(25)

6 Gassmann's Relation; Generalization and Alternative Forms

The most direct way to relate the drained and undrained moduli C_{ijkl} and C^u_{ijkl} is the following. We formally express the stress σ_{ij} as a function of ε_{kl} and m by writing the equation of state (12) or its linearized form (58) as $\sigma_{ij} = \sigma_{ij}(\varepsilon_{kl}, \mu(\varepsilon_{kl}, m))$, assuming the function $\mu = \mu(\varepsilon_{kl}, m)$ to be given, e.g. in the form of the linear relation $\rho_0\mu = -M_\varepsilon \alpha_{ij}\varepsilon_{ij} + M_\varepsilon \Delta m/\rho_0$, obtained by rewriting (59). Differentiating this function one gets

$$\left.\frac{\partial \sigma_{ij}}{\partial \varepsilon_{kl}}\right|_m^0 = C^u_{ijkl} = \left.\frac{\partial \sigma_{ij}}{\partial \varepsilon_{kl}}\right|_\mu^0 + \left.\frac{\partial \sigma_{ij}}{\partial \mu}\right|_{\varepsilon_{kl}}^0 \left.\frac{\partial \mu}{\partial \varepsilon_{kl}}\right|_m^0 = C_{ijkl} + \alpha_{ij}a_{kl},$$

where the derivatives have been evaluated as in (58) and (59); thus we arrive at (64).

Similarly, in establishing the relationship between drained and undrained compliances, one regards the strain as a function $\varepsilon_{ij} = \varepsilon_{ij}(\sigma_{kl}, \mu(\sigma_{kl}, m))$. Making use of the definitions in (65) and (66), one obtains (70)

$$\left.\frac{\partial \varepsilon_{ij}}{\partial \sigma_{kl}}\right|_m^0 = S^u_{ijkl} = \left.\frac{\partial \varepsilon_{ij}}{\partial \sigma_{kl}}\right|_\mu^0 + \left.\frac{\partial \varepsilon_{ij}}{\partial \mu}\right|_{\varepsilon_{kl}}^0 \left.\frac{\partial \mu}{\partial \sigma_{kl}}\right|_m^0 = S_{ijkl} - \beta_{ij}b_{kl}.$$

In applications of poroelasticity theory one often desires to relate drained to undrained properties, or undrained properties for an aqueous pore fluid to undrained properties for hydrocarbon pore fluids.[18] This is the *fluid substitution problem* that continues to be of great interest, in particular to exploration seismologists studying 4-D seismic time-lapse methods (Carcione and Tinivella, 2001). The classical problem considered in this context is essentially the derivation of a relationship between the drained and undrained compliances. This problem is solved by $(70)_1$, but there are a number points to be made here. Consider first the form

$$S_{ijkl} - S^u_{ijkl} = M_\sigma \beta_{ij}\beta_{kl} = \frac{\beta_{ij}\beta_{kl}}{\tilde{S}_\sigma + v_0 c_f} \tag{119}$$

of this result. This tells us that the two compliances can be calculated in terms of each other, once the tensorial coefficient β_{ij} and the scalar

[18] The drained compliances are often referred to as *dry compliances* in this context, in reference to the fact that the pore pressure remains constant in a drained deformation. Note, however, that Biot's notion of a wetted porous medium, as discussed in the above, would be a more appropriate description of the state of a fluid-saturated material that undergoes a drained deformation.

coefficient M_σ are known. Instead of measuring the latter directly, one could also determine $\tilde{S}_\sigma$ and c_f independently. This also suggests a way to account for a change of pore fluid. Although it would thus be desirable to perform a direct measurement of the coefficient $\tilde{S}_\sigma$ by determining the change in pore volume with pore fluid pressure of a sample at constant stress, the difficulty of such a measurement suggests an appeal to expression (91) as an alternative. Proceeding in this way, we also express the components β_{ij} by use of (93) in terms of the components S'_{ijkk}; (119) then assumes the form

$$S_{ijkl} - S^u_{ijkl} = \frac{(S_{ijmm} - S'_{ijmm})(S_{klnn} - S'_{klnn})}{1/K - 1/K'_s + v_0(1/K_f - 1/K''_s)}. \tag{120}$$

This result was first given by Brown and Korringa (1975) and sometimes referred to a 'Brown-Korringa equation' or 'Brown-Korringa relation'. (See also Mavko *et al.* (1998) and Guéguen *et al.* (2004) for discussions of this important relationship.) A particularly important point to bear in mind is that the components S'_{ijkk} are understood to have been measured under $\Pi-$loading conditions in accord with (95), unless of course one is dealing with the special case of a micro-homogenous skeleton, such as a porous single crystal, with known compliances S^s_{ijkl}.

Much earlier, the first result of this kind was derived by Gassmann (1951) for isotropic materials. Gassmann's relation is in fact identical with (106), but by use of (110) and (112) this may be brought into the form

$$\frac{1}{K} - \frac{1}{K_u} = \frac{(1/K - 1/K'_s)^2}{1/K - 1/K'_s + v_0(1/K_f - 1/K''_s)} \tag{121}$$

which represents the isotropic version of (120).

In the interesting more general case of an isotropic unjacketed response, we have

$$S_{ijkl} - S^u_{ijkl} = \frac{[S_{ijmm} - (1/3K'_s)\delta_{ij}][S_{klnn} - (1/3K'_s)\delta_{kl}]}{1/K - 1/K'_s + v_0(1/K_f - 1/K''_s)}. \tag{122}$$

This may often be an acceptable approximation of relation (120), but it is one that must be verified experimentally in each case.

An alternative generalized Gassmann's relation is given by (64), which may be written

$$C^u_{ijkl} - C_{ijkl} = \frac{\alpha_{ij}\alpha_{kl}}{\tilde{S}_\varepsilon + v_0 c_f} \tag{123}$$

and here the possibility of a direct measurement of the Biot coefficient α_{ij} and of the specific storage capacities $\tilde{S}_\varepsilon$ or $S_\varepsilon = \tilde{S}_\varepsilon + v_0 c_f$ seems worth considering. Where this proves impractical, there remains always the option of bringing (123) into the form

$$C^u_{ijkl} - C_{ijkl} = \frac{(\delta_{ij} - C_{ijmn}S'_{mnrr})(\delta_{kl} - C_{klpq}S'_{pqss})}{1/K'_s(1 - K/K'_s) + v_0(1/K_f - 1/K''_s)} \tag{124}$$

by substituting the expressions (94) and (117) for α_{ij} and S_ε. In the case of isotropy (123) reduces to (105) and this can now be expressed as

$$K_u - K = \frac{(1 - K/K'_s)^2}{1/K'_s(1 - K/K'_s) + v_0(1/K_f - 1/K''_s)}, \tag{125}$$

which is just another way of writing (121). Finally, in the special case of an isotropic unjacketed response we may substitute (99) for α_{ij} in (124), to obtain

$$C^u_{ijkl} - C_{ijkl} = \frac{(\delta_{ij} - C_{ijmm}/3K'_s)(\delta_{kl} - C_{klnn}/3K'_s)}{1/K'_s(1 - K/K'_s) + v_0(1/K_f - 1/K''_s)}. \tag{126}$$

Bibliography

Berge, P.A., and J.G. Berryman (1995). Realizability of negative pore compressibility in poroelastic composites. *J. Appl. Mech.* 62, 1053–1062.

Berryman, J.G. (1995). Mixture theories for rock properties. In: *Rock Physics and Phase Relations. A Handbook of Physical Constants*, edited by T.J. Ahrens, Am. Geophys. Union, Washington D.C., pp. 205–228.

Biot, M.A. (1941). General theory of three-dimensional consolidation. *J. Appl. Phys.* 12, 155–164.

Biot, M.A. (1955). Theory of elasticity and consolidation for a porous anisotropic solid. *J. Appl. Phys.* 26, 182–185.

Biot, M.A. (1956a). Thermoelasticity and irreversible thermodynamics. *J. Appl. Phys.* 27, 240–253.

Biot, M.A. (1972). Theory of finite deformation of porous solids. *Indiana Univ. Math. J.* 21(7), 597–620.

Biot, M.A. (1973). Nonlinear and semilinear rheology of porous solids. *J. Geophys. Res.* 78, 4924–4937.

Biot, M.A., and D.G. Willis (1957). The elastic coefficients of the theory of consolidation. *J. Appl. Mech.* 24, 594–601.

Brown, R.J.S., and J. Korringa (1975). On the dependence of the elastic properties of a porous rock on the compressibility of the pore fluid. *Geophysics* 40, 608–616.

Callen, H.B. (1960). *Thermodynamics*, John Wiley, New York.

Carcione, J.M., and U. Tinivella (2001). The seismic response to overpressure: a modelling study based on laboratory, well and seismic data. *Geophysical Prospecting* 49, 523–539.

Carroll, M.M. (1979). An effective stress law for anisotropic elastic deformation. *J. Geophys. Res.* 84, 7510–7512.

Cheng, A.H.D. (1997). Material coefficients of anisotropic poroelasticity. *Int. J. Rock Mech. Min. Sci.* 34, 199–205.

Coussy, O. (2004). *Poromechanics*, John Wiley & Sons Ltd., Chichester UK.

Detournay, E., and A.H.D. Cheng (1993). Fundamentals of poroelasticity. In: *Comprehensive Rock Engineering* Vol. 2, edited by J. A. Hudson, Pergamon Press, Oxford, Chap. 5, pp. 113–171.

Gassmann, F. (1951). Über die Elastizität poröser Medien. *Vierteljahrsschr. Naturforsch. Ges. Zürich* 96, 1–23.

Geertsma, J. (1957a). A remark on the analogy between thermoelasticity and the elasticity of saturated porous media. *J. Mech. Phys. Solids* 6, 13–16.

Geertsma, J. (1957b). The effect of fluid pressure decline on volumetric changes in porous rocks. *Trans. AIME* 210, 331–340.

Geertsma, J. (1966). Problems of rock mechanics in petroleum production engineering. *Proc. 1st Congr. Int. Society of Rock Mechanics*, 585–594.

Green, D.H., and H.F. Wang (1986). Fluid pressure response to undrained compression in saturated sedimentary rocks. *Geophysics* 51, 948–956.

Green, D.H., and H.F. Wang (1990). Specific storage as a poroelastic coefficient. *Water Resources Res.* 26, 1631–37.

Guéguen, Y., L. Dormieux, and M. Boutéca (2004). Fundamentals of Poromechanics. In: *Mechanics of Fluid-Saturated Rocks*, edited by Y. Guéguen and M. Boutéca, Elsevier Academic Press, Burlington MA, pp. 1–54.

Jaeger, J.C., N.G.W. Cook, and R.W. Zimmerman (2007). *Fundamentals of Rock Mechanics* (4th edition), Blackwell Publishing Ltd, Oxford, etc.

Kümpel, H-J. (1991). Poroelasticity: parameters reviewed. *Geophys. J. Int.* 105, 783–799.

Mavko, G., T. Mukerji, and J. Dvorkin (1998). *The Rock Physics Handbook. Tools for Seismic Analysis in Porous Media.* Cambridge U. Press, Cambridge UK.

McTigue, D.F. (1986). Thermoelastic response of fluid-saturated porous rock. *J. Geophys. Res.* 91, 9533–9542.

Nur, A., and J.D. Byerlee (1971). An exact effective stress law for elastic deformation of rock with fluids. *J. Geophys. Res.* 76, 6414–6419.

Nye, J.F. (1957). *Physical Properties of Crystals.* Oxford University Press, Oxford etc., (reprinted in 1979).

Rice, J.R., and M.P. Cleary (1976). Some basic stress diffusion solutions for fluid-saturated elastic porous media with compressible constituents. *Rev. Geophys. Space Phys.* 14, 227–241.

Rudnicki, J.W. (1985). Effect of pore fluid diffusion on deformation and failure of rock. In: *Mechanics of Geomaterials*, edited by E. Bažant, John Wiley & Sons Ltd., New York, Chap. 15, pp. 315–347.

Rudnicki, J.W. (2001). Coupled deformation-diffusion effects in the mechanics of faulting and failure of geomaterials. *Appl. Mech. Rev.* 54, 1–20.

Terzaghi, K., and O.K. Fröhlich (1936). *Theorie der Setzung von Tonschichten*, F. Deutike, Wien.

Thompson, M., and J.R. Willis (1991). A reformulation of the equations of anisotropic poroelasticity. *J. Appl. Mech.* 58, 612–616.

Wang, H.F. (2000). *Theory of Linear Poroelasticity with Applications to Geomechanics and Hydrogeology.* Princeton University Press, Princeton and Oxford.

Weiner, J.H. (1983). *Statistical Mechanics of Elasticity*, John Wiley & Sons, New York.

Zimmerman, R.W., W.H. Somerton, and M.S. King (1986). Compressibility of porous rocks. *J. Geophys. Res.* 91, 12,765–12,777.

Eshelby's Technique for Analyzing Inhomogeneities in Geomechanics

John W. Rudnicki

Department of Civil and Environmental Engineering
and Department of Mechanical Engineering,
Northwestern University, Evanston, Illinois, USA

Abstract This chapter describes Eshelby's technique for determining the stress and strain in regions in an infinite elastic body that undergo a change of size or shape. The technique is also extended to determine the stress and strain in regions of different elastic properties than the surroundings due to loading in the farfield. The chapter also discusses the relation of this technique to singular solutions in elasticity and different integral forms for the solutions. Example applications of the technique include determining the effective stress in a narrow fault zone, the stress and strain in a fluid reservoir and subsidence due to fluid mass injection or withdrawal. The approach of Eshelby is also used as a basis for a brief description of representations of deformation due to slip and seismic source theory.

1 Introduction

The stress and strain in reservoirs, aquifers, intrusions, fault zones, caverns or underground structures can differ significantly from the regional fields because the material response, pore pressure or temperature differs from that of the surrounding material. Often it is of interest to determine the alteration of stress, strain or pore pressure in the inhomogeneity due to remote loading. Examples of this type of problem are the pore pressure change in an aquifer due to barometric or tidal loading or the stress induced in a fault zone by tectonic loading. In other cases, the interest may be in how a change in the inhomogeneity affects other quantities or perturbs the remote field. For example, how does the excavation of an underground storage facility perturb the regional stress? How does injection or withdrawal of fluid mass or a change of pore pressure alter the stress and strain in a subsurface reservoir? Although it is sometimes assumed that the pore pressure change occurs with no change in stress or with the constraint of uniaxial deformation, the propriety of these limiting cases depends on the geometry of the reservoir and the difference in material response from the surroundings.

The well-known solution of Eshelby (1957, 1961) has been widely used to study the effects of inhomogeneities in materials (e.g., Mura, 1987) and provides a relatively simple yet reasonably general approach for exploring these types of problems. In addition, the Eshelby framework can be used as the basis of seismic source theory. Eshelby (1957) originally posed the following problem: Suppose a region of an elastic body, the *inclusion*, undergoes a change of size and shape that could be described by homogeneous strain in the absence of constraint of the surrounding material; what is the resulting state of stress and strain in the inclusion and the surroundings, the *matrix*? Eshelby (1957) did not solve this problem directly but instead used an ingenious series of conceptual steps involving cutting, transforming and re-inserting the inclusion. He showed that for an ellipsoidal inclusion in an infinite elastic body, the resulting stress and strain in the inclusion are uniform. He went on to show that this solution also solves the problem of an elastic inhomogeneity, that is, the perturbation of a remote stress or strain field by an ellipsoidal region of different elastic properties.

This chapter will describe Eshelby's approach and show how the results can be applied to several geomechanical problems. The next section briefly recapitulates some results from elasticity that will be used.

2 Results from Linear Elasticity

In linear elasticity the stress σ_{ij} is related to the infinitesimal strain ε_{kl} by

$$\sigma_{ij} = C_{ijkl}\varepsilon_{kl} \tag{1}$$

Because of the symmetry of σ_{ij} and ε_{kl}, the modulus tensor C_{ijkl} satisfies

$$C_{ijkl} = C_{jikl} = C_{ijlk} \tag{2}$$

In addition, the existence of a strain energy function implies the additional symmetry:

$$C_{ijkl} = C_{klij} \tag{3}$$

For an isotropic material, the modulus tensor is given by

$$C_{ijkl} = G(\delta_{ik}\delta_{jl} + \delta_{il}\delta_{jk}) + (K - 2G/3)\delta_{ij}\delta_{kl} \tag{4}$$

where G is the shear modulus, K is the bulk modulus, and δ_{ij} is the Kronecker delta ($\delta_{ij} = 1$, if $i = j$; $\delta_{ij} = 0$, if $i \neq j$). The combination $K - 2G/3$

is one of the Lamé constants, denoted Λ here; the other Lamé constant is G. Alternatively, the bulk modulus K can be expressed in terms of G and Poisson's ratio ν as $K = 2G(1+\nu)/3(1-2\nu)$. Substituting (4) into (1) yields

$$\sigma_{ij} = 2G\varepsilon_{ij} + (K - 2G/3)\delta_{ij}\epsilon \tag{5}$$

where $\epsilon = \varepsilon_{kk}$ is the volume strain. The strain ε_{ij} is related to the gradient of the displacement u_i by

$$\varepsilon_{ij} = (1/2)\,(u_{i,j} + u_{j,i}) \tag{6}$$

where $(...)_{,i}$ denotes $\partial(...)/\partial x_i$. For quasi-static processes, in which inertia is negligible, the stress σ_{ij} must satisfy the equilibrium equation

$$\sigma_{ij,i} + F_j(\mathbf{x}, t) = 0 \tag{7}$$

where the $F_j(\mathbf{x}, t)$ are the components of the body force per unit volume.

Substituting the strain displacement equation (6) into (5) and the result into equilibrium (7) yields

$$(K + G/3)\epsilon_{,j} + Gu_{j,kk} + F_j(\mathbf{x}, t) = 0 \tag{8}$$

The solution for an arbitrary distribution of body force is given by

$$u_i(\mathbf{x}) = \int_V g_{ij}(\mathbf{x} - \boldsymbol{\xi})F_j(\boldsymbol{\xi}, t)d^3\xi \tag{9}$$

where $g_{ij}(\mathbf{x} - \boldsymbol{\xi})$ is the displacement at $\mathbf{x}$ due to the application of a unit force in the j - direction at the point $\boldsymbol{\xi}$ (Love, 1944; Sokolnikoff, 1956). In other words, the solution of (8) for a singular distribution of body force given by

$$F_j^{Point}(\mathbf{x}, t) = P_j\delta_{DIRAC}(\mathbf{x} - \boldsymbol{\xi}) \tag{10}$$

is

$$u_i(\mathbf{x}) = g_{ij}(\mathbf{x} - \boldsymbol{\xi})P_j \tag{11}$$

where $\delta_{DIRAC}(\mathbf{x})$ is the Dirac delta function defined by the following property

$$\int_V f(\boldsymbol{\xi})\delta_{DIRAC}(\mathbf{x} - \boldsymbol{\xi})d^3\xi = \begin{cases} f(\mathbf{x}) \text{ if } \mathbf{x} \text{ is in } V \\ 0 \text{ if } \mathbf{x} \text{ is not in } V \end{cases} \tag{12}$$

For later reference, note that the strains due to application of a point force are given by

$$\varepsilon_{ij}(\mathbf{x}) = \frac{1}{2}P_k\left(\frac{\partial g_{ik}}{\partial x_j} + \frac{\partial g_{jk}}{\partial x_i}\right)(\mathbf{x} - \boldsymbol{\xi}) \tag{13}$$

and the stresses are given by

$$\sigma_{ij}(\mathbf{x}) = P_m C_{ijkl} \frac{\partial g_{km}}{\partial x_l}(\mathbf{x} - \boldsymbol{\xi}) \tag{14}$$

where we have made use of (2).

In a homogeneous material (the C_{ijkl} do not depend on position) the argument of g_{ij} depends only on the difference $\mathbf{x} - \boldsymbol{\xi}$. Another property of the function $g_{ij}(\mathbf{x} - \boldsymbol{\xi})$ that will be useful later is

$$g_{ij}(\mathbf{x} - \boldsymbol{\xi}) = g_{ji}(\boldsymbol{\xi} - \mathbf{x}) \tag{15}$$

This relation states that the ith component of the displacement at $\mathbf{x}$ due to a unit point force in the jth direction at $\boldsymbol{\xi}$ is equal to the jth component of the displacement at $\boldsymbol{\xi}$ due to a unit point force in the ith direction at $\mathbf{x}$. The result follows from the elastic reciprocal theorem (Sokolnikoff, 1956). Consider the application of two point forces to the body: $\mathbf{P}^A$ at $\boldsymbol{\xi}^A$ and $\mathbf{P}^B$ at $\boldsymbol{\xi}^B$ with corresponding displacement fields $\mathbf{u}^A$ and $\mathbf{u}^B$. The reciprocal theorem requires the work of $\mathbf{P}^A$ on $\mathbf{u}^B$ at $\mathbf{x} = \boldsymbol{\xi}^A$ be equal to the work of $\mathbf{P}^B$ on $\mathbf{u}^A$ at $\mathbf{x} = \boldsymbol{\xi}^B$.

If the elastic body is isotropic and unbounded, then the function g_{ij} is obtained from the Kelvin solution (Love, 1944; Sokolnikoff, 1956) and can be written as

$$g_{ij} = \frac{1}{4\pi G}\left[\frac{\delta_{ij}}{r} + \frac{1}{4(1-\nu)}\frac{\partial^2 r}{\partial x_i \partial x_j}\right] \tag{16}$$

or

$$g_{ij} = \frac{1}{16\pi r G(1-\nu)}\left[(3-4\nu)\delta_{ij} + \frac{(x_i - \xi_i)(x_j - \xi_j)}{r^2}\right] \tag{17}$$

where $r = |\mathbf{x} - \boldsymbol{\xi}|$. The expression for g_{ij} in an elastic half-space can be determined from the solution of Mindlin (1936; 1955). For small values of r, the expression for g_{ij} for the half-space approaches (16) and (17), but it contains additional, non-singular terms that are required to make the traction vanish on the surface of the half-space. Okada (1992) has given a particularly convenient rearrangement of the half-space solution as the sum of a term corresponding to the Kelvin solution (16, 17), an image term, and two depth dependent terms, one of which vanishes on the surface of the half-space.

The point force solution can also be used to generate other singular solutions that will be useful in later discussions. For example, Figure 1a shows the combination of a point force of magnitude M/λ in the x_1 direction at $\boldsymbol{x} = \boldsymbol{\xi} + (\lambda/2)\boldsymbol{e}_1$, where $\boldsymbol{e}_1$ is a unit vector in the x_1 direction, and a force of equal magnitude acting in the opposite direction at $\boldsymbol{x} = \boldsymbol{\xi} - (\lambda/2)\boldsymbol{e}_1$. The displacement is

$$u_i(\boldsymbol{x}) = \frac{M}{\lambda} \{g_{i1}[\boldsymbol{x} - (\boldsymbol{\xi} + (\lambda/2)\boldsymbol{e}_1)] - g_{i1}[\boldsymbol{x} - (\boldsymbol{\xi} - (\lambda/2)\boldsymbol{e}_1)]\} \tag{18}$$

Taking the limit $\lambda \to 0$ yields the displacement due to a force dipole or *double force* of magnitude M in the x_1 direction

$$u_i(\boldsymbol{x}) = M\frac{\partial g_{i1}}{\partial \xi_1}(\boldsymbol{x} - \boldsymbol{\xi}) = -M\frac{\partial g_{i1}}{\partial x_1}(\boldsymbol{x} - \boldsymbol{\xi}) \tag{19}$$

Combining double forces in three perpendicular directions (Figure 1b) yields the displacement field of a *center of dilatation*:

$$u_i(\mathbf{x}) = M\frac{\partial g_{ik}}{\partial \xi_k}(\boldsymbol{x} - \boldsymbol{\xi}) = -M\frac{\partial g_{ik}}{\partial x_k}(\boldsymbol{x} - \boldsymbol{\xi}) \tag{20}$$

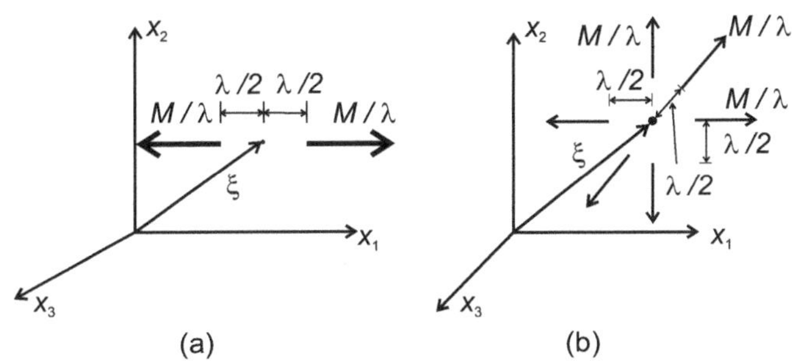

Figure 1. Illustration of (a) a force dipole with axis in the x_1 direction and (b) the combination of three perpendicular force dipoles to form a center-of-dilatation.

Because the forces in Figure 1 are arranged in pairs with a common line of action they combine to give no resultant moment in addition to no resultant force. In contrast, the combination of forces in Figure 2a results in a counter-clockwise moment about the x_3 axis of magnitude M. In the limit $\lambda \to 0$, the displacement field is

$$u_i(\mathbf{x}) = M\frac{\partial g_{i2}}{\partial \xi_1}(\boldsymbol{x} - \boldsymbol{\xi}) = -M\frac{\partial g_{i2}}{\partial x_1}(\boldsymbol{x} - \boldsymbol{\xi}) \tag{21}$$

The combination of point forces in the positive and negative x_1 directions in Figure 2b also results in a couple of magnitude M about the x_3 axis but in the clockwise direction. Combining these two gives the displacement field

$$u_i(\mathbf{x}) = M\left(\frac{\partial g_{i1}}{\partial \xi_2} + \frac{\partial g_{i2}}{\partial \xi_1}\right)(\boldsymbol{x} - \boldsymbol{\xi}) \tag{22}$$

Although this combination of forces is called a *double couple*, in fact, it results in no net force and moment (Similarly, a *double force* results in no net force). Later, Eshelby's result will be used to show that the displacement field is the same as that due to point slip at $\boldsymbol{x} = \boldsymbol{\xi}$. To generalize this

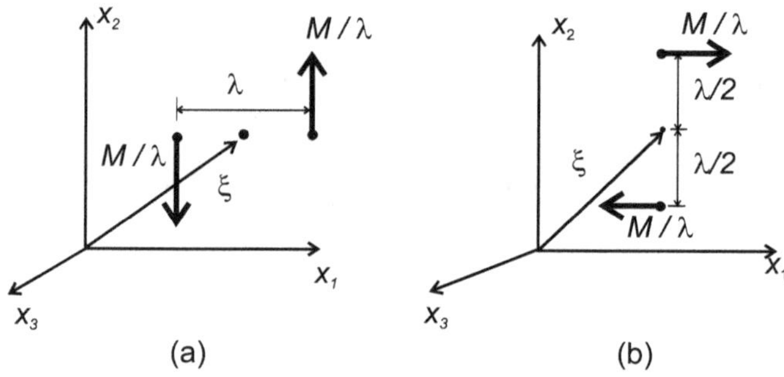

Figure 2. Combination of point forces to give a counterclockwise (a) and clockwise (b) moment about the x_3 axis.

procedure, consider the displacement field due to the sum of a point force $\mathbf{P}$ at $\boldsymbol{\xi} + \boldsymbol{\lambda}/2$ and an oppositely directed point force at $\boldsymbol{\xi} - \boldsymbol{\lambda}/2$ (Figure 3). Taking the limit $|\boldsymbol{\lambda}| \to 0$, but maintaining the product $M_{kl} \to P_k \lambda_l$ finite gives

$$u_i(\mathbf{x}) = M_{kl}\frac{\partial g_{ik}}{\partial \xi_l}(\boldsymbol{x} - \boldsymbol{\xi}) = -M_{kl}\frac{\partial g_{ik}}{\partial x_l}(\boldsymbol{x} - \boldsymbol{\xi}) \tag{23}$$

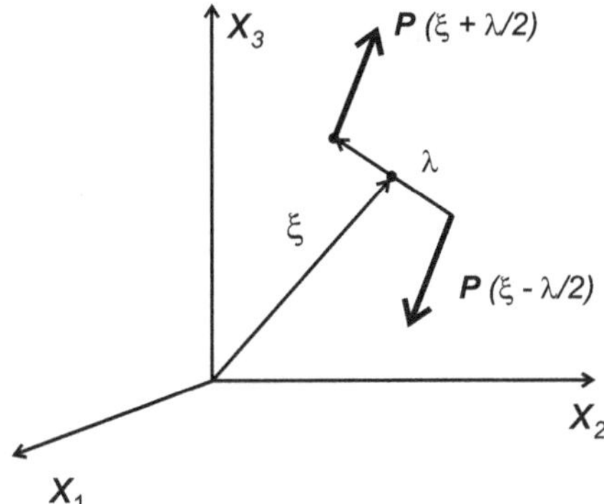

Figure 3. Combination of two point forces to form force dipoles and couples.

If the forces are arranged in pairs so that there is no net force or moment, then $M_{kl} = M_{lk}$ and (23) can be written as

$$u_i(\mathbf{x}) = \frac{1}{2} M_{kl} \left(\frac{\partial g_{ik}}{\partial \xi_l} + \frac{\partial g_{il}}{\partial \xi_k} \right) (\boldsymbol{x} - \boldsymbol{\xi}) \tag{24}$$

The same expression with a minus sign applies if differentiation is with respect to x_l.

3 Results from Eshelby

Now, consider the problem addressed by Eshelby (1957): The elastic field caused by a change of size and shape of a region, the *inclusion*, in an isotropic elastic body. The treatment here draws heavily on the exposition of seismic source theory by Rice (1980). He treats the dynamic problem so that the inertia terms are included in (7) and the function corresponding to g_{ij} in (11) also depends on the difference between the current time and the time at which the point force is introduced. The development here is for the quasi-static case which Rice (1980) discusses as a special case of the dynamic result.

The sequence of operations used by Eshelby (1957) to solve this problem is shown in Figure 4. First, imagine cutting the region V, the inclusion, free from the elastic body. Since the body is stress-free, this can be done without introducing any body forces or tractions.

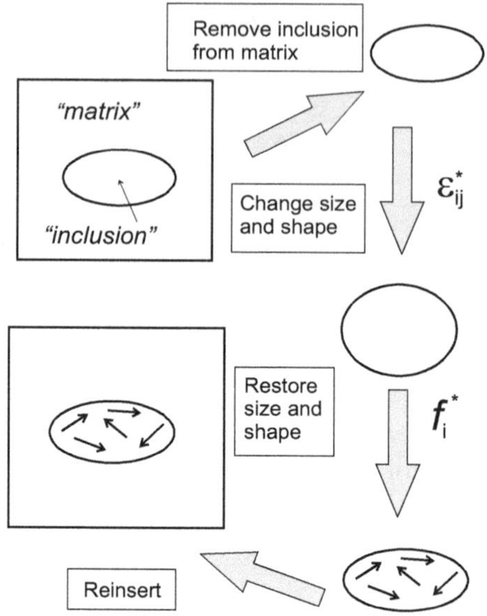

Figure 4. Sequence of cutting and welding operations used by Eshelby (1957) to determine the elastic field of an inclusion.

Next, alter the size and shape of the region V by application of the strain field $\varepsilon^*_{kl}(\boldsymbol{\xi})$ (Eshelby (1957) considered this strain to be uniform, a case we will discuss in more detail below.) Physical examples of $\varepsilon^*_{kl}(\xi)$ include a phase change, shrinkage due to cooling of a region of a thermoelastic solid, or expansion due to injection of fluid into a region of a poroelastic solid. Consequently, $\varepsilon^*_{kl}(\boldsymbol{\xi})$ is called the *transformation strain.*

The change of size and shape described by the transformation strain occurs without stress and, hence, both the inclusion and the elastic body remain free from stress. Now, restore the region V to its original size and shape by the application of a stress field

$$\sigma^*_{ij}(\boldsymbol{\xi}) = -C_{ijkl}\varepsilon^*_{kl}(\boldsymbol{\xi}) \tag{25}$$

In order for this stress field to be in equilibrium, it is necessary to apply simultaneously a body force field $F^*_j(\boldsymbol{\xi})$ that satisfies (7):

$$F^*_j(\boldsymbol{\xi}) = \frac{\partial}{\partial \xi_k}\left[C_{kjlm}\varepsilon^*_{lm}(\boldsymbol{\xi})\right] \tag{26}$$

At this point, the strain is zero everywhere and the stress vanishes outside V. Inside V, the stress is $\sigma^*_{ij}(\boldsymbol{\xi})$ and there is an extraneous body force field given by $F^*_j(\boldsymbol{\xi})$.

The unwanted body force distribution can be removed by applying its negative and using (9) to calculate the resulting displacement field:

$$u_i(\mathbf{x}) = -\int_V g_{ij}(\mathbf{x}-\boldsymbol{\xi}) \frac{\partial}{\partial \xi_k} \left[C_{kjlm}\varepsilon^*_{lm}(\boldsymbol{\xi})\right] d^3\xi \tag{27}$$

The desired solution is the sum of the displacement, strain and stress fields resulting from (27) and those that existed before the removal of the body force field. Because the displacement is zero everywhere prior to the removal of the body force, (27) gives the displacement for the problem originally posed. The strain field can be obtained by differentiation of (27). The stress outside V can be calculated from the strain but, inside V, it is necessary to add (25). Thus, the complete stress field is

$$\sigma_{ij}(\mathbf{x}) = \begin{cases} C_{ijkl}\left[\varepsilon_{kl}(\mathbf{x}) - \varepsilon^*_{kl}(\mathbf{x})\right], & \text{inside } V \\ C_{ijkl}\varepsilon_{kl}(\mathbf{x}), & \text{outside } V \end{cases} \tag{28}$$

Because the transformation strain $\varepsilon^*_{kl}(\boldsymbol{\xi})$ vanishes outside the finite region V, it is convenient to rewrite (27) as the sum of two terms

$$u_i(\mathbf{x}) = -\int_V \frac{\partial}{\partial \xi_k}\left[g_{ij}(\mathbf{x}-\boldsymbol{\xi})C_{kjlm}\varepsilon^*_{lm}(\boldsymbol{\xi})\right] d^3\xi + \int_V \frac{\partial g_{ij}}{\partial \xi_k}(\mathbf{x}-\boldsymbol{\xi})\left[C_{kjlm}\varepsilon^*_{lm}(\boldsymbol{\xi})\right] d^3\xi \tag{29}$$

Using the divergence theorem to evaluate the first term on a surface just outside V, where $\varepsilon^*_{kl} = 0$, leaves

$$u_i(\mathbf{x}) = \int_V \frac{\partial g_{ij}}{\partial \xi_k}(\mathbf{x}-\boldsymbol{\xi})\left[C_{kjlm}\varepsilon^*_{lm}(\boldsymbol{\xi})\right] d^3\xi \tag{30}$$

Because of the symmetry (2) of C_{kjlm}, the integrand can here be written in the form (24) with $M_{kl} = C_{klmn}\varepsilon^*_{mn}$. Thus, the displacement is due to a distribution of double forces without moment over the volume V. Another intepretation of (30) is obtained by using (2), (3) and the result from the reciprocal theorem (15) to rewrite (30) as follows:

$$u_i(\mathbf{x}) = \int_V \left[C_{lmjk}\frac{\partial g_{ji}}{\partial \xi_k}(\boldsymbol{\xi}-\mathbf{x})\right] \varepsilon^*_{lm}(\boldsymbol{\xi}) d^3\xi \tag{31}$$

Comparison with (14) reveals that the term in brackets in the integrand is the stress at $\boldsymbol{\xi}$ due to a unit point force in the ith direction at $\mathbf{x}$:

$$\sigma_{lm}^{(i)}(\boldsymbol{\xi}-\mathbf{x}) = C_{lmjk}\frac{\partial g_{ji}}{\partial \xi_k}(\boldsymbol{\xi}-\mathbf{x}) \tag{32}$$

Thus, (31) can be written as

$$u_i(\mathbf{x}) = \int_V \sigma_{lm}^{(i)}(\boldsymbol{\xi}-\mathbf{x})\varepsilon_{lm}^*(\boldsymbol{\xi})d^3\xi \tag{33}$$

Consequently, the result (31) can be interpreted as following from the equality of work expressions:

$$\int_V P_i\delta_{DIRAC}(\boldsymbol{\xi}-\mathbf{x})u_i(\boldsymbol{\xi})d^3\xi = \int_V P_i\sigma_{lm}^{(i)}(\boldsymbol{\xi}-\mathbf{x})\varepsilon_{lm}^*(\boldsymbol{\xi})d^3\xi \tag{34}$$

The left term is the work of the point force P_i at $\mathbf{x}$ on the displacement at $\boldsymbol{\xi}$ due to the transformation strain $\varepsilon_{lm}^*(\boldsymbol{\xi})$. The right side is the work of the stress at $\boldsymbol{\xi}$ due to the point force P_i at $\mathbf{x}$ on the transformation strain $\varepsilon_{lm}^*(\boldsymbol{\xi})$. The left side is defined for the entire body, not just the volume V and since the transformation strain vanishes outside V, both integrals can be extended to the entire body. Thus, the symmetry (3) and the constraint of the reciprocal theorem on the form of g_{ij} (15) lead to an interpretation of the Eshelby result (30) in terms of the reciprocal theorem.

3.1 Thin Inclusions

Often reservoirs or fault zones are very thin, i.e., they have a thickness much less than their lateral extent. Consequently, it is useful to specialize (30) to this case. For simplicity, consider a narrow volume V with midsurface Σ perpendicular to the $x_3(\xi_3)$ direction. Because the thickness of the zone is assumed to be small, the displacement near $\xi_3 = 0$ can be written as

$$u_i^*(\boldsymbol{\xi}) = \frac{b_i(\xi_1,\xi_2)}{h(\xi_1,\xi_2)}\xi_3 \tag{35}$$

where $h(\xi_1,\xi_2)$ is the thickness and the $b_i(\xi_1,\xi_2)$ are the net relative displacements across the zone. The strains are

$$\varepsilon_{33}^* = \frac{b_3(\xi_1,\xi_2)}{h(\xi_1,\xi_2)} \tag{36a}$$

$$\varepsilon_{3\alpha}^* = \frac{1}{2}\left(\frac{b_\alpha(\xi_1,\xi_2)}{h(\xi_1,\xi_2)} + \xi_3\frac{\partial}{\partial\xi_\alpha}\left[\frac{b_3(\xi_1,\xi_2)}{h(\xi_1,\xi_2)}\right]\right) \tag{36b}$$

$$\varepsilon_{\alpha\beta}^* = \frac{1}{2}\xi_3\left(\frac{\partial}{\partial\xi_\alpha}\left[\frac{b_\beta(\xi_1,\xi_2)}{h(\xi_1,\xi_2)}\right] + \frac{\partial}{\partial\xi_\beta}\left[\frac{b_\alpha(\xi_1,\xi_2)}{h(\xi_1,\xi_2)}\right]\right) \tag{36c}$$

where $(\alpha, \beta) = (1, 2)$. Substituting into (30), expanding other terms in the integrand about $\xi_3 = 0$, integrating ξ_3 from $-h/2$ to $+h/2$ and taking the limit $h \to 0$ yields

$$u_i(\mathbf{x}) = \int_\Sigma \frac{\partial g_{ij}}{\partial \xi_k}(\mathbf{x} - \boldsymbol{\xi}) C_{kj3m} b_m(\boldsymbol{\xi}) d\xi_1 d\xi_2 \tag{37}$$

Generalizing the result to a surface Σ with unit normal $\mathbf{n}$ (pointing from the $(-)$ to the $(+)$ side of Σ) yields

$$u_i(\mathbf{x}) = \int_\Sigma \frac{\partial g_{ij}}{\partial \xi_k}(\mathbf{x} - \boldsymbol{\xi}) C_{kjlm} \frac{1}{2}(n_l b_m + n_m b_l)(\boldsymbol{\xi}) d\xi_1 d\xi_2 \tag{38}$$

The same expression can be obtained directly from (30) by writing the transformation strain as

$$\varepsilon_{ij}^*(\boldsymbol{\xi}) = \frac{1}{2}(n_i b_j + n_j b_i)(\boldsymbol{\xi}) \delta_{DIRAC}(\Sigma) \tag{39}$$

where $\delta_{DIRAC}(\Sigma)$ is the singular Dirac delta function converting the domain of the integral from the volume V to the surface Σ.

Comparison of the integrand in (38) with (24) shows that the displacement is due to a distribution of double forces without moment with strengths $C_{kjnm} n_n b_m$ over the surface Σ. The same manipulations that lead from (30) to (31) and use of (2), (3), and (15) allow us to rewrite (38) as

$$u_i(\mathbf{x}) = \int_\Sigma \left[C_{lmkj} \frac{\partial g_{ji}}{\partial \xi_k}(\boldsymbol{\xi} - \mathbf{x}) \right] \frac{1}{2}(n_l b_m + n_m b_l)(\boldsymbol{\xi}) d\xi_1 d\xi_2 \tag{40}$$

Using (32) to again identify the term in square brackets as the stress at $\boldsymbol{\xi}$ due to a unit point force in the ith direction at $\mathbf{x}$ yields the Volterra formula for the displacement field due to a distribution of dislocations (displacement discontinuities, b_k) on Σ (Steketee, 1958):

$$u_i(\mathbf{x}) = \int_\Sigma n_l \sigma_{lm}^{(i)}(\boldsymbol{\xi} - \mathbf{x}) b_m(\boldsymbol{\xi}) d\xi_1 d\xi_2 \tag{41}$$

As noted by Rice (1980), this expression is typically derived beginning with an application of the reciprocal theorem as expressed in (34).

Specializing (38) to an isotropic material (5) gives

$$\begin{aligned} u_i(\mathbf{x}) \quad = \quad & \int_\Sigma \left[\Lambda n_m b_m(\boldsymbol{\xi}) \frac{\partial g_{ik}}{\partial \xi_k}(\mathbf{x} - \boldsymbol{\xi}) \right. \\ & \left. + G n_l b_m(\boldsymbol{\xi}) (\frac{\partial g_{il}}{\partial \xi_m} + \frac{\partial g_{im}}{\partial \xi_l})(\mathbf{x} - \boldsymbol{\xi}) \right] d\xi_1 d\xi_2 \end{aligned} \tag{42}$$

If the relative displacement is purely normal to the plane of the zone, opening or overlap, $b_m(\boldsymbol{\xi}) = b_{\text{normal}}(\boldsymbol{\xi})n_m$ and (42) becomes

$$\begin{aligned} u_i(\mathbf{x}) &= \int_\Sigma b_{\text{normal}}(\boldsymbol{\xi}) \left[\Lambda \frac{\partial g_{ik}}{\partial \xi_k}(\mathbf{x} - \boldsymbol{\xi}) \right. \\ & \left. + 2G n_l \, n_m \left(\frac{\partial g_{il}}{\partial \xi_m} \right) (\mathbf{x} - \boldsymbol{\xi}) \right] d\xi_1 d\xi_2 \end{aligned} \tag{43}$$

Thus, the source on Σ is the combination of a center-of-dilatation, weighted by Λ, and a force dipole in the direction $\mathbf{n}$ weighted by $2G$. If the displacement is evaluated far from the surface Σ (compared with a characteristic length of Σ) then the term in square brackets in the integrand can be evaluated at any point on Σ and taken outside the integral. The remaining integration is equal to the average relative normal displacement over Σ multiplied by the area of Σ. This justifies identification of the quantity in square brackets as a point *uniaxial strain* (or, less accurately, a *tensile*) source. If the relative displacement is perpendicular to the normal, that is, pure slip, $n_m b_m = 0$, and (42) reduces to

$$u_i(\mathbf{x}) = \int_\Sigma G n_l b_m(\boldsymbol{\xi}) \left(\frac{\partial g_{il}}{\partial \xi_m} + \frac{\partial g_{im}}{\partial \xi_l} \right) (\mathbf{x} - \boldsymbol{\xi}) d\xi_1 d\xi_2 \tag{44}$$

Evaluation of the displacement far from Σ leads to a double couple source (without moment) of strength equal to the product of G, the average slip and the area of Σ.

3.2 Volumetric Transformation Strain

For later reference, we record the forms of the preceding expressions when the transformation strain is purely volumetric

$$\varepsilon^*_{lm}(\boldsymbol{\xi}) = \frac{1}{3} \epsilon^*(\boldsymbol{\xi}) \delta_{lm} \tag{45}$$

If the material is isotropic (5), then

$$C_{kjlm} \varepsilon^*_{lm}(\boldsymbol{\xi}) = K \epsilon^*(\boldsymbol{\xi}) \delta_{kj} \tag{46}$$

With (46), (30) becomes

$$u_i(\mathbf{x}) = \int_V K \epsilon^*(\boldsymbol{\xi}) \frac{\partial g_{ik}}{\partial \xi_k}(\mathbf{x} - \boldsymbol{\xi}) d^3\xi \tag{47}$$

Comparison of (47) with (20) demonstrates that the displacement is the result of a distribution of centers of dilatation of strength $K\epsilon^*(\boldsymbol{\xi})$ over the volume V. When the transformation strain results from a temperature change, (47) reflects the *nuclei of strain* method introduced by Goodier (Timoshenko and Goodier, 1970).

Alternatively, substituting (45) into (33) yields

$$u_i(\mathbf{x}) = \int_V \frac{1}{3}\sigma_{kk}^{(i)}(\boldsymbol{\xi} - \mathbf{x})\epsilon^*(\boldsymbol{\xi})d^3\xi \tag{48}$$

This is a form of the Maysel integral representation (Maysel, 1941) (also see Nowacki (1986)) discussed by Lehner *et al.* (2005). The result was used by Maysel to determine the displacement field due to a temperature change in a region, but is adapted by Lehner *et al.* (2005) to solve an analogous problem of poroelasticity corresponding to fluid withdrawal or injection from a region. Although (48) and (47) are equivalent, Lehner *et al.* (2005) note that the former has certain advantages for computations in inhomogeneous media. They obtain (48) from an adaption of the reciprocal theorem to a poroelastic solid and its application analogous to (34). Although obtained by a different route here, the reciprocal theorem is used via the property (15).

3.3 Uniform Transformation Strain

In the problem originally considered by Eshelby (1957), the transformation strain $\varepsilon_{kl}^*(\boldsymbol{\xi})$ was taken as uniform, ε_{ij}^T. Consequently, the size and shape of the inclusion (after transformation and before re-insertion into the matrix) can be restored by a traction applied to the surface:

$$t_j^T = -n_k C_{kjlm}\varepsilon_{lm}^T \tag{49}$$

Thus, the displacement caused by the removal of this tractions is that due to a distribution of point forces over the boundary of the inclusion

$$u_i(\mathbf{x}) = -C_{kjlm}\varepsilon_{lm}^T \int_S n_k g_{ij}(\mathbf{x} - \boldsymbol{\xi})d^2\xi \tag{50}$$

The strain, obtained by differentiation, is

$$\varepsilon_{rs}(\mathbf{x}) = -C_{kjlm}\varepsilon_{lm}^T \int_S n_k \frac{1}{2}\left[\frac{\partial g_{rj}}{\partial x_s} + \frac{\partial g_{sj}}{\partial x_r}\right](\mathbf{x} - \boldsymbol{\xi})d^2\xi \tag{51}$$

The same result is obtained by recognizing that the derivative appearing in the integrand of (27) is zero everywhere except on the boundary of V. On the boundary, the derivative is singular as the $\varepsilon_{ij}^*(\boldsymbol{\xi})$ drops from its uniform value, ε_{ij}^T inside V to zero outside and can be represented formally as

$$\frac{\partial}{\partial \xi_\alpha}\left\{C_{\alpha jkl}\varepsilon_{kl}^T\right\} = -n_\alpha C_{\alpha jkl}\varepsilon_{kl}^T \delta_{DIRAC}(S) \tag{52}$$

where δ_{DIRAC} is again the Dirac delta function, S is the surface bounding V, and the negative sign appears because the strain decreases upon crossing S in the direction of the outward normal with components n_i. Substitution into (27) changes the volume integral to a surface integral and reduces this expression to (50).

If the body is unbounded and isotropic and the volume V is an ellipsoid, then Eshelby (1957) showed that the integrals in (51) can be expressed explicitly in terms of elliptic integrals. The resulting constrained strain in the inclusion (i.e., that which results when the transformation strain occurs under the constraint of the surrounding material) is uniform and can be expressed as

$$\varepsilon_{mn} = S_{mnkl}\varepsilon^T_{kl} \tag{53}$$

The array of factors S_{ijkl} depend only on Poisson's ratio and the geometry of the ellipsoid. (Eshelby (1957) also notes that the strain is uniform in an inclusion in an anisotropic elastic solid.) General expressions for the S_{ijkl} are given by Eshelby (1957) and by Mura (1987). In axes coinciding with the principal axes of the ellipsoid, the only nonzero entries are of the form S_{iijj} or S_{ijij} and the array is symmetric with respect to interchange of the first and last two indices, but not with respect to interchange of the first pair and last pair. Rudnicki (1977) gives expressions and tabulates some results for the S_{ijkl} for oblate and prolate axisymmetric ellipsoids, Rudnicki (2002a) gives extensive graphical results for oblate axisymmetric ellipsoids, and Mura (1987) gives expressions for axisymmetric and cylindrical shapes.

3.4 Ellipsoidal Inhomogeneity

For uniform transformation strain of an ellipsoidal region in an infinite elastic body, the constrained strain is given by (53). The stress inside the inclusion, from (28), is

$$\sigma_{ij} = C_{ijkl}\left[\varepsilon_{kl} - \varepsilon^T_{kl}\right] \tag{54}$$

Eliminating the transformation strain between (53) and (54) yields

$$\sigma_{ij} = C_{ijkl}\left[\delta_{kn}\delta_{lm} - S^{-1}_{klmn}\right]\varepsilon_{mn} \tag{55}$$

The relation (55) reflects the constraint of the elastic material exterior to the inclusion. Stated differently, if an ellipsoidal cavity is loaded by a traction $n_i\sigma_{ij}$ derived from the uniform stress σ_{ij}, then the resulting strain of the cavity boundary is compatible with the uniform strain ε_{mn} where σ_{ij} and ε_{mn} are related by (55). Because the strain of the interior is uniform, the displacement of the boundary is $\varepsilon_{mn}x_n$.

If the solid is loaded by farfield stresses σ_{ij}^{∞} and strains $\varepsilon_{ij}^{\infty}$ related by

$$\sigma_{ij}^{\infty} = C_{ijkl}\varepsilon_{kl}^{\infty} \tag{56}$$

then ε_{ij} and σ_{ij} in (54) and (55) should be replaced by the differences $\varepsilon_{ij}^{I} - \varepsilon_{ij}^{\infty}$ and $\sigma_{ij}^{I} - \sigma_{ij}^{\infty}$ where the superscript 'I' denotes the value in the inclusion. Thus, (55) becomes

$$\sigma_{ij}^{I} - \sigma_{ij}^{\infty} = C_{ijkl}\left[\delta_{kn}\delta_{lm} - S_{klmn}^{-1}\right]\left(\varepsilon_{mn}^{I} - \varepsilon_{mn}^{\infty}\right) \tag{57}$$

Note that σ_{ij}^{I} and ε_{mn}^{I} are not necessarily related by (1) and that (57) expresses the constraint of the elastic material surrounding the ellipsoidal region. Consequently, the elastic inclusion can be replaced any homogeneous material with the stress related to the strain by $\sigma_{ij}^{I} = F_{ij}\left[\varepsilon_{mn}^{I}\right]$ and the stress and strain at the boundary are required to satisfy (57). In other words, the transformation strain ε_{kl}^{T}, which has been eliminated from (57) can always be chosen so that the constitutive relation of the material in the ellipsoidal region can be satisfied. For example, (57) solves the problem of the deformation of an ellipsoidal cavity in an infinite elastic solid loaded in the farfield by σ_{ij}^{∞}. In this case, $\sigma_{ij}^{I} = 0$ since the boundary of the cavity is traction-free and the displacement of the boundary is given by $u_m = \varepsilon_{mn}^{I}x_n$ (and $\varepsilon_{mn}^{\infty}$ is related to σ_{ij}^{∞} by (56)). As another example, Rudnicki (1977) has used an incremental version of (57) with a pressure-sensitive, dilatant, strain-softening elastic-plastic relation in the ellipsoidal region as a model of the evolution of a fault zone due to the slow increase of far-field strains. In particular, he examines the relation between the point at which conditions for localization of deformation are met in the ellipsoidal region and the occurrence of an inertial instability corresponding to the onset of a seismic event.

3.5 Poroelastic Inhomogeneity

Rudnicki (2002a) has examined in detail the case in which the ellipsoidal region is poroelastic with a constitutive relation of the form

$$\sigma_{ij}^{I} = C_{ijkl}^{I}\varepsilon_{kl}^{I} + \varsigma^{I}p^{I}\delta_{ij} \tag{58}$$

where p is the alteration in pore fluid pressure, C_{ijkl}^{I} is the matrix of drained (corresponding to zero change in pore pressure) moduli and ς^{I} is a porous media constant. This constant is often denoted α and is equal to $1 - K_I/K_s'$, where K_I is the bulk modulus of the porous inclusion material and K_s' is an additional bulk modulus. Under circumstances discussed by Rice and Cleary (1976) and Wang (2000), K_s' can be identified with the bulk modulus of the solid constituents, K_s, and, more generally, is of the same order. The identification K_s' with K_s is equivalent to the requirement that the relative change in pore volume fraction is equal to the bulk dilatation of the solid phase (see eqn. (85) in Chapter 1). Although temperature changes are not considered here, they could be included by appending the term $K_I \alpha_s' \Delta\theta \delta_{ij}$ to the right hand side of (58), where α_s' is the cubical thermal expansion coefficient and $\Delta\theta$ is the temperature change. Alternatively, (58) could be expressed in terms of the undrained moduli (corresponding to the response for zero change in pore fluid mass) and the alteration of fluid mass. Because the ellipsoidal region must be homogeneous, the pore pressure must be uniform. This will be a good approximation if the ellipsoidal region is much more permeable than the surroundings. In addition, the relation (57) is based on purely elastic response in the region surrounding the ellipsoid and, hence, does not account for exchange of fluid with the ellipsoidal region. Thus, the region surrounding the ellipsoid is assumed to deform under either drained or undrained conditons. These assumptions are, however, reasonable for application to reservoirs, aquifers, caverns, and, possibly, fault zones. Rudnicki (2002a) has given extensive graphical results for the effects of geometry and material mismatch on the stresses induced by fluid injection or withdrawal.

It is important to keep in mind that even if the inhomogeneity is ellipsoidal, but embedded in a bounded region, the constrained strain will not be uniform. In this case, the simple result (57) relating differences in stress and strain will not apply. For many of the applications mentioned in the preceding paragraph, the idealization of an inhomogeneity embedded in a half-space (rather than an infinite space) will be a better approximation. Nevertheless, the results for the infinite space are attractively simple and will be a good approximation as long as the inhomogeneity is not too close to the surface of the half-space.

If the ellipsoidal region is poroelastic, then (58) and (56) can be used to eliminate either the stresses or strains from (57). The result is a relation between the stress (or strain) and pore pressure (or fluid mass change) of the ellipsoidal region in terms of the farfield stress (or strain). For isotropic response, the relation between the strains is

$$\begin{aligned}\epsilon^I_{mn} + S_{mnkk}\, e^I \left(\frac{K_I}{K} - \frac{G_I}{G}\right) + \left(\frac{G_I}{G} - 1\right) S_{mnkl}\, \varepsilon^I_{kl} \\ = \epsilon^\infty_{mn} - \frac{1}{3}\frac{\varsigma_I p^I}{K} S_{mnkk} \qquad (59)\end{aligned}$$

where $\epsilon^I = \epsilon^I_{kk}$, G_I and K_I are the shear modulus and drained bulk modulus of the inclusion and the same symbols without subscripts are the moduli of the surrounding material. If the strains are eliminated, the result can be written as the following two equations:

$$s^I_{mn} + \left(\frac{G_I}{G} - 1\right)\left\{S_{mnkl} - \frac{S_{ppkl}\left[(K_I/K - 1)\,S_{mnqq} + \delta_{mn}\right]}{3\left[1 + \alpha\,(K_I/K - 1)\right]}\right\} s^I_{kl}$$
$$= \frac{G_I}{G} s^\infty_{mn} - \frac{2G_I}{3K}\frac{\left[S_{mnkk} - \alpha\delta_{mn}\right]}{\left[1 + \alpha\,(K_I/K - 1)\right]}\left(\varsigma_I p^I + \left(\frac{K_I}{K} - 1\right)\right) \tag{60}$$

and

$$\sigma^I\left[1 + \alpha\left(\frac{K_I}{K} - 1\right)\right] + \left(\frac{G_I}{G} - 1\right)\frac{K_I}{2G_I} S_{mnkl} s^I_{kl}$$
$$= \frac{K_I}{K}\sigma^\infty + (1 - \alpha)\varsigma_I p^I \tag{61}$$

where the stress has been separated into spherical and deviatoric parts $\sigma_{ij} = s_{ij} + (\sigma/3)\delta_{ij}$ and $3\alpha = S_{ppqq} = (1 + \nu_\infty)/(1 - \nu_\infty)$.

4 Applications

4.1 Effective Stress in a Narrow Fault Zone

Cocco and Rice (2002) have examined whether the pore pressure induced in a narrow, flat fault zone for undrained conditions (no change in fluid mass content) is determined by the mean stress or the normal stress. Here, we reproduce their result from the above analysis in the limit of a flat, ellipsoidal inhomogeneity with normal in the x_3 direction. In this limit, the only non-zero values of the array S_{ijkl} are

$$S_{3333} = 1,\ S_{1313} = S_{2323} = 1/2$$
$$S_{3311} = S_{3322} = \nu_\infty/(1 - \nu_\infty)$$

(Rudnicki (2002a) gives the expansions including the first order term in the aspect ratio of the inclusion.) With these values it is possible to determine from (57) or from the limits of (59), (60) and (61) that $\sigma^I_{ij} = \sigma^\infty_{ij}$ whenever i or j is 3 and that $\varepsilon^I_{ij} = \varepsilon^\infty_{ij}$ when neither i nor j is 3. Cocco and Rice (2002) deduce these results directly from equilibrium and compatibility. The requirement that $\sigma^I_{33} = \sigma^\infty_{33}$, which appears as the third of (19) in Rudnicki (2002a) can be written in the more compact form

$$M_I\epsilon^I_{33} = M_\infty\epsilon^\infty_{33} + (\epsilon^\infty_{11} + \epsilon^\infty_{22})(\Lambda_\infty - \Lambda_I) + \varsigma_I p^I \tag{62}$$

where $M = 2G(1 - \nu)/(1 - 2\nu)$ is the elastic modulus for one-dimensional strain and $\Lambda = 2G\nu/(1 - 2\nu)$ is a Lamé constant. Specializing (61) for a flat axisymmetric ellipsoid, setting $p^I = 0$, consistent with taking the elastic moduli of the inhomogeneity to be undrained values, and using $\sigma^I_{33} = \sigma^\infty_{33}$ yield the mean stress in the inhomogeneity $\sigma^I_{kk}/3$ in terms of the remote normal stress σ^∞_{33} and the remote mean stress $\sigma^\infty_{kk}/3$. For undrained conditions, the pore pressure induced in the fault zone is equal to $-B_I\sigma^I_{kk}/3$, where B_I is Skempton's coefficient in the fault zone (Cocco and Rice, 2002):

$$p^I = -B_I\frac{K_I}{M_I}\left\{\frac{G_I}{G_\infty}\frac{M_\infty}{K_\infty}\sigma^\infty_{kk}/3 + \left[1 - \frac{G_I}{G_\infty}\right]\sigma^\infty_{33}\right\} \tag{63}$$

As noted by Cocco and Rice (2002), the pore pressure in the fault zone is determined by the normal stress σ^∞_{33} if the shear modulus of the fault zone is much less than that of the surroundings ($G_I \ll G_\infty$). On the other hand, the pore pressure is determined by the mean normal stress $\sigma^\infty_{kk}/3$, if the shear moduli are nearly the same ($G_I \simeq G_\infty$). Cocco and Rice (2002) also note that for anisotropy so pronounced that fluid pumping causes strain only in the x_3 direction when the fault is held at constant stress, the pore pressure is determined solely by the normal stress.

4.2 Fluid Mass Injection or Withdrawal

Determining the stress and strain changes due to the injection or withdrawal of fluid mass is one of the fundamental problems of aquifer or reservoir management. Rudnicki (2002a) has used the Eshelby approach to examine in detail the effects of geometry and elastic mismatch on the stresses and strains induced by pressure or fluid mass changes in an ellipsoidal region embedded in an unbounded linear elastic solid. He assumes that the fluid mass or pressure changes occur on a time scale much shorter than that for changes of regional stress or strain and, hence, σ^∞_{ij} and ε^∞_{ij} may be taken as zero. For an axisymmetric ellipsoidal region with the x_3 direction distinguished, Rudnicki (2002a) gives the ratio of lateral strain change to axial strain change, $R = \Delta\varepsilon^I_{11}/\Delta\varepsilon^I_{33}$, due to a pressure change Δp^I, as

$$R = \frac{3\alpha - S_{33kk} + (G_I/G - 1)\left\{3\alpha S_{3333} - S_{33kk}S_{pp33}\right\}}{2S_{33kk} + (G_I/G - 1)\left\{3\alpha S_{3333} - S_{33kk}S_{pp33}\right\}} \tag{64}$$

where the expressions for the relevant S_{ijkl} in this case are given by

$$\begin{aligned}
\alpha &= (1/3)S_{kkjj} = (1+\nu_\infty)/3(1-\nu_\infty) \\
S_{kk33} &= 1 - \frac{(1-2\nu_\infty)}{(1-\nu_\infty)} I(e) \\
S_{3333} &= \frac{1}{2}\left(1 + S_{kk33}\right) - \frac{e^2(2-3I(e))}{2(1-\nu_\infty)(1-e^2)} \\
S_{33kk} &= 3\alpha\left(1 - I(e)\right)
\end{aligned}$$

$e = c/a$ is the aspect ratio (ratio of shorter principal axis to longer), and

$$I(e) = \frac{e}{(1-e^2)^{3/2}} \left\{ \arccos(e) - e(1-e^2)^{1/2} \right\} \tag{65}$$

The expression (65) corrects Equation (22) of Rudnicki (2002a), which has e^2 in the numerator of the first term outside the bracket. (I am extremely grateful to Colin Sayers, Schlumberger, Houston, USA for tracking down this misprint.) Figure 5 (after Figure 4 of Rudnicki (2002a)) plots the strain ratio R against the aspect ratio of the reservoir, e, for three values of the ratio of the reservoir shear modulus to matrix shear modulus G_I/G, 0.5, 1.0, and 2.0. (The plot is for $\nu = 0.2$, but the dependence on ν is weak). Although the strain state induced in the reservoir is often assumed to be uniaxial, Teufel *et al.* (1991) have questioned the validity of this assumption. The plot demonstrates that uniaxial strain ($R = 0$) is a good assumption for thin reservoirs ($e \ll 1$), but the approach to uniaxial strain is slower for larger values of G_I/G.

The axial strain change can be written as (Rudnicki, 2002a, eqns. (26) and (27))

$$\Delta\varepsilon^I_{33} = -\varsigma_I p^I E_3/K_I \tag{66}$$

where

$$E_3 = \frac{\alpha(K_I/K)}{(1+2R)\left[1+\alpha\left(K_I/K-1\right)\right] + \left(G_I/G-1\right)(1-R)(S_{kk33}-\alpha)} \tag{67}$$

Equations (64), (66) and (67) can be used to calculate the slope of the stress path caused by injection or withdrawal. In particular, Figure 6 shows a diagram of q versus s where $q = (\bar{\sigma}_3 - \bar{\sigma}_1)/2$, $s = (\bar{\sigma}_3 + \bar{\sigma}_1)/2$, σ_3 and σ_1 are the maximum and minimum principal stresses (positive in compression) and $\bar{\sigma}_i = \sigma_i - p$ is the form of the effective stress appropriate for failure (Rice, 1977; Paterson and Wong, 2005). If injection simply increased the pore pressure without altering the other stress components, injection would cause the stress point to move horizontally to the left ($\Delta s < 0$) by an amount

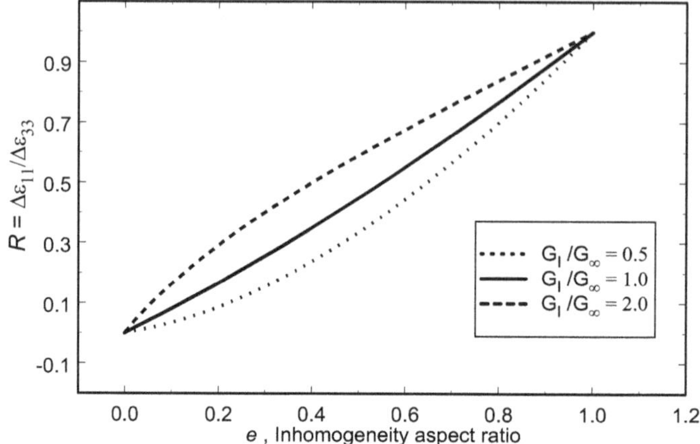

Figure 5. Ratio of lateral to axial strain induced by a pressure change in a reservoir embedded in an infinite elastic solid (after Rudnicki, 2002a, Fig. 4).

equal to the change in pore pressure. Similarly, withdrawal would cause the stress point to move horizontally to the right ($\Delta s > 0$). Because the change of pore pressure generally does alter the other stress components, $\Delta q \neq 0$ (and, also, Δs is not simply equal to the change in pore pressure). The slope $\Delta q/\Delta s$ is given by (Rudnicki, 2002a, eqn. (31))

$$\frac{\Delta q}{\Delta s} = \frac{\varsigma_I E_3 (G_I/K_I)(1-R)}{1-\varsigma_I \left\{1-\left[3E_3/2(1+\nu_I)\right]\left[1+R(1+2\nu_I)\right]\right\}} \tag{68}$$

In the limit of a very narrow zone ($e \ll 1$ and $R = 0$) this ratio becomes

$$\frac{\Delta q}{\Delta s} = \left[\frac{2(1-\nu_I)}{\varsigma_I(1-2\nu_I)} - 1\right]^{-1} \tag{69}$$

(This expression corrects Eqn. (32) of Rudnicki (2002a) which contains several misprints. I am grateful to F. Mulders, Delft Technical University for identifying the errors in this expression.) Because Δq and Δs have the same sign, injection causes the stress point to move down and to the left and withdrawal causes it to move up and to the right. Figure 7 (after Fig. 10 of Rudnicki (2002a)) plots values of the slope $\Delta q/\Delta s$ against the Poisson's ratio of the reservoir for various aspect ratios, $G_I/G = 1.0$ and $\varsigma_I = 0.9$. The line for $e = 0$ shows the slopes that occur from uniaxial strain (69), but, as Figure 7 shows, the slopes are smaller for reservoirs of finite aspect ratio. The values of all the slopes decrease with increasing values of the Poisson's ratio of the inhomogeneity.

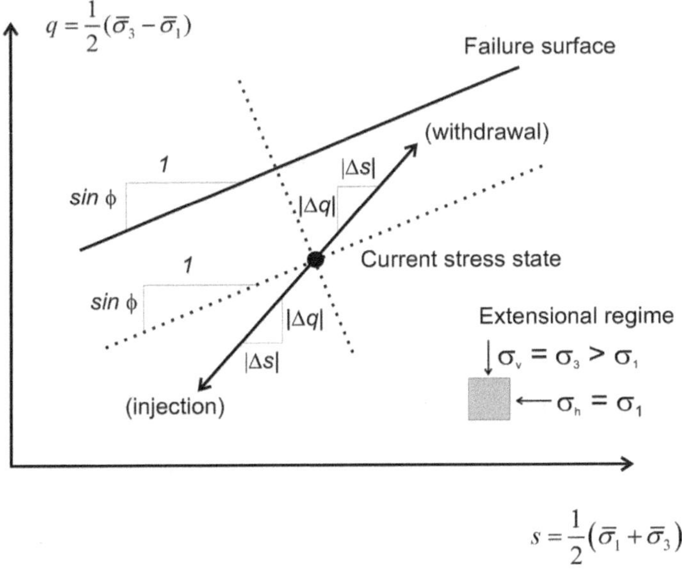

Figure 6. Illustration of stress paths and approach to failure by fluid injection or withdrawal.

Mulders (2003) has shown that the results of the ellipsoidal inhomogeneity model agree well with finite element calculations and has used the finite element method to explore more complex geometries. Holt et al. (2004) have used the model to examine the effects of stress path, in particular, deviations from the condition of uniaxial strain, on compaction and pore compressibility within the poroelastic approximation. Holt et al. (2004) use a discrete element model to explore non-elastic effects.

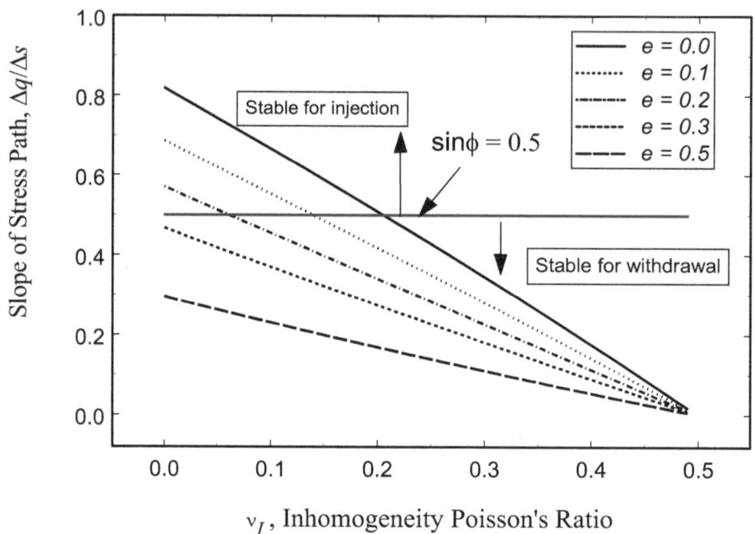

Figure 7. Plots of the stress path $\Delta q/\Delta s$ against the inhomogeneity aspect ratio ν_I for various inclusion aspect ratios. For a friction angle $\phi = 30\,^\circ$ and a normal stress regime, values above (below) the horizontal line, corresponding to $\sin\phi = 0.5$, move the stress away from (toward) the failure line.

If a failure criterion is adopted, then the expressions (68) or (69) can be used to assess in a simple way whether the pressure changes due to injection or withdrawal will move the stress state of the reservoir closer to or further from failure. Figure 6 illustrates the situation for a linear Mohr-Coulomb failure envelope. For an inhomogeneity with short axis in the vertical direction, a normal stress state with a vertical stress $\sigma_{\mathrm{v}} = \sigma_3$ that exceeds the horizontal stress $\sigma_{\mathrm{h}} = \sigma_1$ corresponds to positive q and plots in the first quadrant. Since (68) shows that Δq has the same sign as Δs, withdrawal will cause the stress state will move closer to the failure envelope if

$$\frac{\Delta q}{\Delta s} > \sin\phi \tag{70}$$

where ϕ is the friction angle. According to this condition, if $\phi = 30°$, values of the slope above the line corresponding to $\sin\phi = 0.5$ shown in Figure 7 would be unstable for withdrawal in the sense of moving the stress toward the failure line; values below would be unstable for injection. If the current stress state is typical of a compressional environment, that is, the horizontal stress $\sigma_{\text{h}} = \sigma_1$ exceeds the vertical stress $\sigma_{\text{v}} = \sigma_3$ (again, for an inhomogeneity with short axis in the vertical direction), then $q < 0$ for the current stress and the current stress state plots in the fourth quadrant. The slope of the failure line in this quadrant is $-\sin\phi$. Consequently, withdrawal is always stable and injection is always unstable. Of course, this analysis assumes linear poroelasticity and is of little consequence if the current stress is far from the failure envelope by comparison to the change in stress state. In addition, it is well to reemphasize that the results assume the region of injection or withdrawal is embedded in an infinite space.

4.3 Subsidence Due to Fluid Mass Injection or Withdrawal

In a series of papers, Geertsma (1966, 1973,a) addressed the problem of land subsidence due to fluid withdrawal from subsurface reservoirs using a Green's function approach based on an analogy of poroelasticity and thermoelasticity (Geertsma, 1957). The approach of Geertsma can be reconstructed by substituting (58), specialized to an isotropic material, into (7). Because the gradient of the pore pressure now appears as a body force, the representation (9) can be used to calculate the displacement, strains and stress. Because the gradient of the pore pressure appears and the alteration in pore pressure vanishes outside a finite volume, the same manipulations used in going from (27) to (30) can be used to express the displacement as an integration of a distribution of centers of dilatation given by the pore pressure change. In the corresponding thermo-elastic result, the temperature plays the role of the pore pressure and for many materials of practical interest, the constant corresponding to ς is nearly unity. Segall (1985) and Mossop and Segall (1997) have also considered subsidence due to fluid withdrawal using the approach of Geertsma and, in addition, have calculated the induced stress changes and used them to assess effects on seismicity (Segall, 1985, 1989, 1992; Segall *et al.*, 1994). Both Geertsma and Segall work out results for a flat, tabular reservoir geometry and use the g_{ij} for an infinite half space (Mindlin, 1936, 1955; Mindlin and Cheng, 1950; Okada, 1992).

Not surprisingly, the same results(though limited to an infinite space) are obtained from the Eshelby framework. Specializing (58) to the isotropic case and taking the trace yields

$$\sigma = K\epsilon - \varsigma p \tag{71}$$

where $\sigma = (1/3)\sigma_{kk}$ is the mean stress, $\epsilon = \varepsilon_{kk}$ is the volume strain, and the superscript I denoting the inclusion material has been dropped. Setting the stress to zero defines the *stress-free strain* or *transformation strain* ϵ^* resulting from an alteration of pore pressure

$$\epsilon^* = \varsigma p/K \tag{72}$$

Substituting into (47) gives the same result as Geertsma and shows that the displacement is the result of a distribution of centers-of-dilatation. (As noted above, $\varsigma \approx 1$ in Geertsma's analysis based on the thermoelastic analogy.)

Recently, Walsh (2002) has reexamined the problem using Eshelby's approach. In particular, he begins by assuming that the reservoir undergoes a change of fluid mass, and then procedes to follow the Eshelby procedure of transformation, application of tractions to restore the size and shape, re-insertion into the matrix, and removal of tractions. As shown by Walsh (2002) and discussed by Rudnicki (2002b), this procedure arrives at the same result as Geertsma's analysis if the reservoir (inclusion) is poroelastic and the changes in pore pressure in each step are accounted for. The result, in either case is that the displacements are given by (47) although a novel aspect of Walsh (2002)'s analysis is that he uses the reciprocal theorem to evaluate expressions for the strain of the reservoir and the surface subsidence.

Lehner *et al.* (2005) generalize Walsh (2002)'s use of the reciprocal theorem to establish a reciprocal theorem for a poroelastic solid. They use it to obtain an integral representation for the displacement due to fluid injection or withdrawal and recognize its analogy to Maysel (1941)'s solution of the corresponding thermoelastic problem. As noted by Lehner *et al.* (2005) and mentioned earlier in this chapter, the result is equivalent to that obtained by Goodier's (Timoshenko and Goodier, 1970) *nucleus of strain* method which has been the predominant one for subsidence problems (and, also, for a variety of other geophysical problems such as fault slip and magma injection (Segall, 2009)). Lehner *et al.* (2005) explain, however, that the Maysel representation can be applied more easily to problems of more general geometries and inhomogeneous materials. In particular, it offers significant computational advantages.

The computational advantages of the Lehner *et al.* (2005) Maysel formulation can be understood by comparison of (47) and (48). In (47), the term $\partial g_{ik}(\mathbf{x}-\boldsymbol{\xi})/\partial\xi_k$ is the solution for the displacement at $\mathbf{x}$ due to a unit center-of-dilatation at $\boldsymbol{\xi}$, where $\boldsymbol{\xi}$ is the variable to be integrated over the volume for which the transformation strain is non-zero. Consequently, it is necessary to re-compute this term whenever the distribution of transformation strain or the volume of integration changes. In contrast, the term $\sigma_{kk}^{(i)}(\boldsymbol{\xi})/3$ in (48) is the mean stress at $\boldsymbol{\xi}$ due to a point force at $\mathbf{x}$. Since the purpose is to calculate the displacement at $\mathbf{x}$ this term need be computed only once for a given geometry regardless of changes in the distribution of transformation strain or the volume over which it occurs.

4.4 Deformation Due to Slip

As noted earlier, the treatment of Eshelby's analysis follows the treatment of Rice (1980) of seismic source theory. Conversely, the Eshelby framework also provides a convenient entrée to seismic source theory (Aki and Richards, 1980; Rice, 1980; Kostrov and Das, 1988) and here we briefly recapitulate some of the results of Rice (1980). As noted in connection with (30), the displacement can be regarded as due to the integration of force dipoles and double couples of density $C_{ijkl}\varepsilon_{kl}^*$. This motivates the terminology of *seismic moment density tensor* for the negative of the right hand side of (25)

$$m_{ij}(\mathbf{x}, t) = C_{ijkl}\varepsilon_{kl}^* \tag{73}$$

where the transformation strain ε_{kl}^* is now a function of time in addition to position. Equation (26) then expresses the body force equivalence of seismic sources. The necessary Green's function, denoted $G_{ij}(\mathbf{x}-\boldsymbol{\xi}, t-t')$, is the solution to (8) with the inertia terms included on the right hand side for an instantaneous (delta function time dependence) body force given by (10). The dynamic version of (30) includes an additional superposition over time

$$u_i(\mathbf{x}, t) = \int_V \int_{-\infty}^{t} \frac{\partial G_{ij}}{\partial \xi_k}(\mathbf{x}-\boldsymbol{\xi}, t-t')m_{kj}(\boldsymbol{\xi}, t')dt'd^3\xi \tag{74}$$

As in the static case, the derivatives $\partial G_{ij}/\partial\xi_k$ represent force dipoles and couples. Because of (2), the seismic moment tensor is also symmetric. As a result (74) can be written as

$$u_i(\mathbf{x}, t) = \int_V \int_{-\infty}^{t} \frac{1}{2}\left\{\frac{\partial G_{ij}}{\partial \xi_k}(\mathbf{x}-\boldsymbol{\xi}, t-t') + \frac{\partial G_{ik}}{\partial \xi_j}(\mathbf{x}-\boldsymbol{\xi}, t-t')\right\} \tag{75}$$
$$\times\, m_{kj}(\boldsymbol{\xi}, t')dt'd^3\xi$$

and the sources represented in the integrand are double forces and double couples without moment.

Although the form of $G_{ij}(\mathbf{x}-\boldsymbol{\xi}, t-t')$ is known for an infinite, isotropic elastic solid and other simple cases, the Earth is strongly nonhomogeneous, anisotropic and, at least partially non-elastic. Indeed, the challenge of seismology, stated succintly, is to use observations of the left side of (75) at points on the surface of the Earth to infer information about its composition and the nature of the source generating the displacement. A further complication is that the instrument measuring the displacement may be accurate only in a range of magnitudes and frequencies.

Despite the absence of sources with a net moment in the representation, the following quantity is defined as the *seismic moment tensor*:

$$M_{ij}(t) = \int_V m_{ij}(\boldsymbol{\xi}, t) d^3\xi \tag{76}$$

The seismic moment tensor plays an important role in observations of earthquakes both from seismograms and from measurements of surface displacements after earthquakes. If the material is isotropic and the source is confined to a surface, e.g., a fault, then (39) and (73) lead to

$$M_{ij}(t) = \int_\Sigma \{\Lambda n_m b_m(\boldsymbol{\xi}, t)\delta_{ij} + G\left[n_i b_j(\boldsymbol{\xi}, t) + n_j b_i(\boldsymbol{\xi}, t)\right]\} d\xi_1 d\xi_2 \tag{77}$$

For shear faulting (no opening) $n_m b_m = 0$; if, in addition, the fault surface is flat with normal in the x_3 direction and slip $\Delta u = b_1(\xi_1, \xi_2, t)$ only in the x_1 direction, the only nonzero components of $M_{ij}(t)$ are $M_{13} = M_{31} = M$ where

$$M(t) = G \int_\Sigma \Delta u(\xi_1, \xi_2, t) d\xi_1 d\xi_2 \tag{78}$$

In an isotropic material, the displacement field (75) is the sum of $u_i^d(\mathbf{x}, t)$ and $u_i^s(\mathbf{x}, t)$, the contributions from the dilatational and shear waves, respectively. Expressions for $u_i^d(\mathbf{x}, t)$ and $u_i^s(\mathbf{x}, t)$ in the farfield are

$$u_i^d = \frac{\gamma_i \gamma_j \gamma_k}{4\pi\rho r c_d^3} \int_{V(\boldsymbol{\xi})} \dot{m}_{kj}(\boldsymbol{\xi}, t - r/c_d + \boldsymbol{\gamma}\cdot\boldsymbol{\xi}/c_d) d^3\xi) \tag{79}$$

and

$$\begin{aligned} u_i^s = & \frac{1}{4\pi\rho r c_s^3}\left[\frac{1}{2}(\delta_{ij}\gamma_k + \delta_{ik}\gamma_j) - \gamma_i\gamma_j\gamma_k\right] \\ & \times \int_{V(\boldsymbol{\xi})} \dot{m}_{kj}(\boldsymbol{\xi}, t - r/c_s + \boldsymbol{\gamma}\cdot\boldsymbol{\xi}/c_s) d^3\xi \end{aligned} \tag{80}$$

where r is the distance from a point in the source region to the observation point, $\boldsymbol{\gamma} = (\mathbf{x} - \boldsymbol{\xi})/r$, ρ is the density, $c_d = \sqrt{(\Lambda + 2G)/\rho}$ is the dilatational wave speed, $c_s = \sqrt{G/\rho}$ is the shear wave speed, and the superposed dot denotes the time derivative. The farfield approximation is accurate for distances r that are much larger than the characteristic source dimension and $2\pi c/\varpi$, the wavelength of radiation associated with the frequency ϖ. As discussed by Rice (1980), Kostrov (1970) has shown that the source can be approximated as a point if the time differences $\boldsymbol{\gamma}\cdot\boldsymbol{\xi}/c$ between different points of the source region can be neglected (This imposes a limit on the upper frequency for which the representation is accurate). In this case, the integrals in (79) and (80) define the time dependence of the moment and the displacements can be written as

$$u_i^d = \frac{\gamma_i\gamma_j\gamma_k}{4\pi\rho r_0 c_d^3}\dot{M}_{kj}(\boldsymbol{\xi}, t - r_0/c_d) \tag{81}$$

$$u_i^s = \frac{1}{4\pi\rho r_0 c_s^3}\left[\frac{1}{2}(\delta_{ij}\gamma_k + \delta_{ik}\gamma_j) - \gamma_i\gamma_j\gamma_k\right]\dot{M}_{kj}(\boldsymbol{\xi}, t - r_0/c_s) \tag{82}$$

The Fourier transforms of the displacements (81) and (82) are proportional to terms of the form

$$\exp(-\iota\varpi r_0/c)\left[\iota\varpi\tilde{M}_{kj}(\varpi)\right]$$

where $\tilde{M}_{kj}(\varpi)$ is the Fourier transform of $M_{kj}(t)$. As explained by Rice (1980), the short term, seismic motion of an earthquake can be regarded as essentially complete for long periods. Consequently, $\tilde{M}_{kj}(\varpi)$ can be regarded as the transform of a constant, $\tilde{M}_{kj}(\varpi) \approx M_{kj}(t_r)/\iota\varpi$, where t_r is the time at which rupture is completed. Therefore, the Fourier spectrum of the displacement becomes flat at low frequency and its magnitude is characterized by the moment at the end of rupture. For a double couple source, the magnitude is the product of the shear modulus, the average slip and the area of the slip surface (44).

Acknowledgement

Thanks to the organizers Yves Leroy and Florian Lehner for inviting me to take part in the CISM course and to Jean Rudnicki for preparing some of the figures. I am also grateful to Florian for bringing to my attention the Maysel representation. The U.S. Department of Energy, Office of Basic Energy Sciences provided partial financial support of some of my own research reported in this chapter.

Bibliography

Aki, K. and P.G. Richards (1980). *Quantitative Seismology* Vol. 1. W. H. Freeman and Company, San Francisco.

Cocco, M. and J.R. Rice (2002). Pore pressure and poroelasticity effects in Coulomb stress analysis of earthquake interactions. *J. Geophys. Res.* 107(B2), 10.1029/2000JB000138.

Eshelby, J.D. (1957). The determination of the elastic field of an ellipsoidal inclusion and related problems. *Proceedings of the Royal Society of London A* 241, 376–396.

Eshelby, J.D. (1961). Elastic inclusions and inhomogeneities. In *Progress in Solid Mechanics* Vol. 2, edited by I.N. Sneddon and R. Hill, pp. 87–140, North-Holland, Amsterdam.

Geertsma, J. (1957). A remark on the analogy between thermoelasticity and the elasticity of saturated porous media. *J. Mech. Phys. Solids* 6, 13–16.

Geertsma, J. (1966). Problems of rock mechanics in petroleum engineering. In *Proc. 1st Congr. Int. Soc. Rock Mechanics* Vol. 1, 585–594, National Laboratory of Civil Engineering, Lisbon.

Geertsma, J. (1973). Land subsidence above compacting oil and gas reservoirs. *J. Petrol. Technology* 25, 734–744.

Geertsma, J. (1973a). A basic theory of subsidence due to reservoir compaction: The homogeneous case. *Verh. Kon. Ned. Geol. Mijnbouwk. Gen.* 28, 43–62.

Holt, R. M., O. Flornes, L. Li, and E. Fjær (2004). Consequences of depletion-induced stress changes on reservoir compaction and recovery. In *Rock Mechanics Across Borders and Disciplines, Proceedings of Gulf Rocks 2004, the 6th North American Rock Mechanics Symposium*, Paper No. 589.

Kostrov, B.V. (1970). The theory of the focus for tectonic earthquakes. *Izv. Earth Physics* 4, 84–101.

Kostrov, B.V. and S. Das (1988). *Principles of Earthquake Source Mechanics.* Cambridge Monographs on Mechanics and Applied Mathematics. Cambridge University Press, Cambridge.

Lehner, F. K., J. K. Knoglinger, and F. D. Fischer (2005). Use of a Maysel integral representation for solving poroelastic inclusion problems. In *Thermal Stresses 2005*, Proc. 6th Int. Congr. on Thermal Stresses, May 26–29, 2005, Vienna, Austria, edited by F. Ziegler, R. Heuer, and C. Adam, pp. 77-80, Schriftenreihe der TU Wien.

Love, A.E.H. (1944). *A Treatise on the Mathematical Theory of Elasticity*, 4th edition. Dover Publications, New York. Republication of 4th (1927) edition published by Cambridge University Press.

Maysel, V.M. (1941). A generalization of the Betti-Maxwell theorem to the case of thermal stresses and some of its applications (in Russian). *Dokl. Acad. Nauk. USSR* 30, 115–118.

Mindlin, R.D. (1955). Force at a point in the interior of a semi-infinite solid. In *Proc. 1st Midwestern Conf. Solid Mechanics*, pp. 56–59.

Mindlin, R.D. (1936). Force at a point in the interior of a semi-infinite solid. *Physics* 7, 195–202.

Mindlin, R.D. and D.H. Cheng (1950). Nuclei of strain in the semi-infinite solid. *J. Appl. Physics* 21, 926–933.

Mossop, A. and P. Segall (1997). Subsidence at The Geysers geothermal field, N. California from a comparison of GPS and leveling surveys. *Geophys. Res. Lett.* 24(14), 1839–1842.

Mulders, F.M.M. (2003). *Modelling of Stress Development and Fault Slip in and Around a Producing Gas Reservoir*. PhD thesis, Delft University of Technology, Delft, The Netherlands.

Mura, T. (1987). *Micromechanics of Defects in Solids*, 2nd revised edition. Kluwer Academic Publishers, Norwell.

Nowacki, W. (1986). *Thermoelasticity*, 2nd edition. PWN-Polish Scientific Publishers, Warsaw; Pergamon Press, Oxford.

Okada, Y. (1992). Internal deformation due to shear and tensile faults in a half-space. *Bull. Seismological Society of America* 82, 1018–1040.

Paterson, M. and T.-f. Wong (2005). *Experimental Rock Deformation: The Brittle Field*, 2nd edition. Springer-Verlag, Berlin etc.

Rice, J.R. (1977). Pore pressure effects in inelastic constitutive formulations for fissured rock masses. In *Advances in Civil Engineering through Engineering Mechanics*, pp. 360–363. American Society of Civil Engineers, New York.

Rice, J.R. (1980). The mechanics of earthquake rupture. In *Physics of the Earth's Interior* (International School of Physics "Enrico Fermi", Course 78, 1979) edited by A.M. Dziewonski and E. Boschi, pp. 555–649, North Holland Publishing Company, Amsterdam.

Rice, J.R. and M.P. Cleary (1976). Some basic stress diffusion solutions for fluid-saturated elastic porous media with compressible constituents. *Reviews of Geophysics and Space Physics* 14, 227–241.

Rudnicki, J.W. (1977). The inception of faulting in a rock mass with a weakened zone. *J. Geophys. Res.* 82, 844–854.

Rudnicki, J.W. (2002a). Alteration of regional stress by reservoirs and other inhomogeneities: Stabilizing or destabilizing? In *Proc. 9th Int. Congr. Rock Mechanics* Vol. 3, Paris, Aug. 25-29, 1999, edited by G. Vouille and P. Berest, pp. 1629–1637, Balkema, Rotterdam.

Rudnicki, J.W. (2002b). Eshelby transformations, pore pressure and fluid mass changes, and subsidence. In *Poromechanics II, Proc. 2nd Biot Conf. on Poromechanics*, Grenoble, Aug. 26-28, edited by J.-L. Auriault, C. Geindreau, P. Royer, J.-F. Bloch, C. Boutin, and J. Lewandowska, pp. 307–312, Balkema, Rotterdam.

Segall, P. (1985). Stress and subsidence resulting from subsurface fluid withdrawal in the epicentral region of the 1983 Coalinga earthquake. *J. Geophys. Res.* 90, 6801–6816.

Segall, P. (1989). Earthquakes triggered by fluid extraction. *Geology* 17, 942–946.

Segall, P. (1992). Induced stresses due to fluid extraction from axisymmetric reservoirs. *Pure and Applied Geophysics* 139, 535–560.

Segall, P., J.-R. Grasso, and A. Mossop (1994). Poroelastic stressing and induced seismicity near the Lacq gas field, southwestern France. *J. Geophys. Res.* 99, 15,423–15,438.

Segall, P. (2010). *Earthquake and Volcano Deformation.* Princeton University Press, Princeton.

Sokolnikoff, S. (1956). *Mathematical Theory of Elasticity*, 2nd edition, McGraw-Hill, New York.

Steketee, J. A. (1958). Some geophysical applications of the elasticity theory of dislocations. *Canadian Journal of Physics* 36, 1168–1198.

Teufel, L.W., D. W. Rhett, and H. E. Farrell (1991). Effect of reservoir depletion and pore pressure drawdown on in situ stress and deformation in the Ekofisk field, North Sea. In *Proc. 32nd U.S. Symposium on Rock Mechanics*, pp. 63–72, Balkema, Rotterdam.

Timoshenko, S.P. and J. N. Goodier. *Theory of Elasticity*, 3rd edition, McGraw-Hill, New York.

Walsh, J. B. (2002). Subsidence above a planar reservoir. *J. Geophys. Res.* 107(B9), doi:10.1029/2001JB000606.

Wang, H. F. (2000). *Theory of Linear Poroelasticity with Applications to Geomechanics and Hydrology*, Princeton University Press, Princeton.

Effective Elastic Properties of Cracked Rocks - An Overview

Yves Guéguen[†] and Mark Kachanov[‡]

[†] Laboratoire de Géologie, CNRS, Ecole Normale Supérieure, Paris, France

[‡] Department of Mechanical Engineering, Tufts University, Boston MA, USA

AbstractUpper crustal rocks contain cracks of diverse sizes, shapes and orientations. Predicting their influence on the effective elastic properties of a rock poses a challenging problem of considerable interest, in particular to seismologists. In geophysics, theoretical work on this problem is usually complicated by the presence of pore fluids or when the defect-free *matrix* of a rock is elastically anisotropic, as in the case of shales for example. The first challenge arising in this context is the identification of microstructure-sentive parameters in terms of which the effective elastic constants are to be expressed. Simple parameters such as volume fraction or crack density (as usually defined) may not suffice for capturing the way in which the complex microstructure of a rock determines its effective elasticity. Defect interactions present another challenge and a number of approximate theoretical schemes have been designed to deal with this problem, but their applicability is not always clear. This chapter offers a critical assessment of these questions, putting the emphasis on microcracks as microstructural elements that often have a dominant effect on the overall elasticity. The issues of matrix anisotropy and frequency effects are examined. Because elastic wave velocities carry microstructural information, it is possible and of great interest to extract from them information on pore and crack content, as well as on the presence and properties of pore fluids. Problems related to multiple defects have been discussed in a broader context of mechanics of materials for over half a century. One of the goals of the present work is to bridge a gap in this area between geophysics and the general mechanics of materials. The former does not always utilize the results of the latter; the latter often neglects specific complexities of the former.

1 Introduction

This chapter deals with theoretical explanations of the effect of cracks and pores in upper crustal rocks on their effective elastic properties. Since the latter determine the wave velocities, a quantification of this effect is of considerable interest, for example in interpreting 4D seismics data in connection with monitoring reservoirs, underground repositories, or faults. On one hand, the influence of multiple defects on effective properties of elastic solids constitutes one of the main topics in a broader field of mechanics of materials, where it has been discussed for half a century. Substantial progress made in this field has not always been utilized in geophysics. On the other hand, the specific complexity encountered in geophysics, such as highly *irregular* defect geometries, frequency effects, etc., have not been sufficiently addressed in the mechanics of materials. While the present work will hopefully contribute to closing this gap, its limited length will demand a focus on the interests of the authors.

The rocks of the Earth's upper crust (down to 10-15 km depth) are typically porous to some extent and often contain cracks, fractures, and faults as 'defects' on very different length scales. By way of a coarse subdivision, one may distinguish microscale defects from large-scale defects as follows.

Microscale defects depend on the rock type. Sedimentary rocks may have porosities up to some 40% with pore sizes in the range from 10 nm to 100 μm. Pore shapes are highly irregular. In clastic sediments they may tend to be concave when pores represent an intergranular space (Figure 1). The overall physical properties strongly depend on the pore space geometry (Zimmerman, 1991; Guéguen and Palciauskas, 1994). Next to pores, cracks are common microscale defects. While the porosity associated with these cracks is usually very small, microcracks play a major role in determining the macroscopic elastic properties of a rock.

The porosity of igneous rocks is usually very small. In granites for instance it is less than 5%. The main defects are then microcracks (Figure 2). The case of basalts is very different, however; porosities can be much higher (large pores can exist due to gas exsolution) while cracks may also be present.

The fact that microcracks can alter the effective elastic properties of a rock substantially, in spite of their very small pore volume, was noted by Simmons and Brace (1965) and Walsh (1965a). The effects of microcracks on fracture, dilatancy, and permeability were first addressed by Brace (1961), Brace and Bombolakis (1963), Brace *et al.* (1966, 1968), and Zoback and Byerlee (1975). A detailed review of crack microstructures was given by Simmons and Richter (1976). Microcrack propagation under compression was modeled by several authors (for a review of 2-D models, see Lehner and Kachanov, 1996). Microcrack propagation under compression was experimentally examined in more realistic 3-D setting by Dyskin *et al.* (1999). An analysis of damage, fractures, and crack effects on elastic properties and

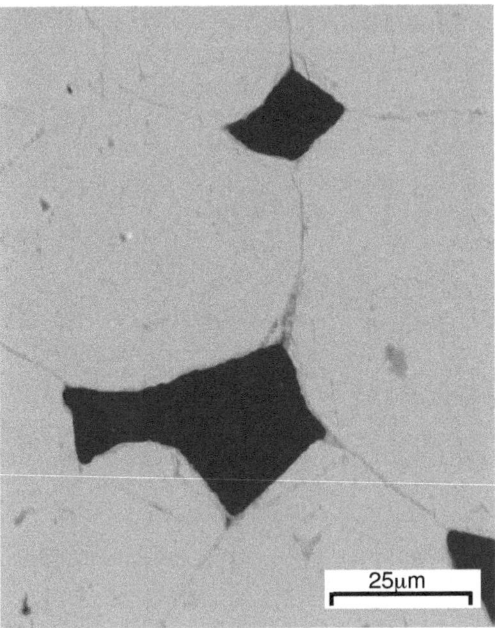

Figure 1. SEM image of pores in Fontainebleau sandstone at 17% Porosity. Quartz grains and SiO_2 cement. Note slightly cracked grain boundaries. (Courtesy J. Fortin)

permeability in rocks can be found in Dresen and Guéguen (2004). Three microcracks types can be distinguished: (1) intergranular microcracks that follow grain boundaries, (2) intragranular microcracks, within a grain, that often emanate from a pore or a grain boundary and (3) transgranular microcracks that run across one or several grains. Grain sizes and therefore microcrack sizes range typically from 1 μm to 1 mm.

Large-scale defects in the range from 1 m upwards may be cracks, joints or fractures (Jaeger *et al.*, 2007). A system of fractures may constitute a fault zone, but faults and fault zones or fracture zones can extend over hundreds to several thousand km in length, depending on their geological type. Where fault zones comprise families of approximately parallel fractures, their effective elasticity at this scale may become of interest in the context of exploration seismology. Similarly to microcracks, these large-scale features can be modeled as displacement discontinuities, although the

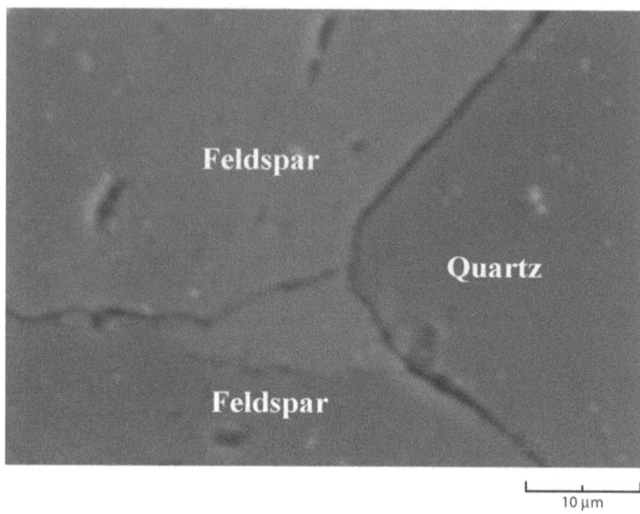

Figure 2. Cracked boundary and intragranular crack (SEM) in Westerly granite (from Nasseri *et al.*, 2006).

response of the latter is controlled by different microstructural parameters, i.e., primarily by contacts between fracture faces (for quantitative models see Kachanov *et al.*, 2010).

Importantly, at both the micro- and the macroscale, the problem of relating effective elastic properties to relevant structural features is often of a similar kind. This chapter will nevertheless focus on microscale effects, for which laboratory-scale experimental data are available.

As has already been noted, the relationship *microstructure* ↔ *overall property* represents a classical theme in the mechanics of heterogeneous materials. In the general area of solid mechanics/materials science, it began to attract attention in the 1950's, with classical works of Mackenzie (1950), Eshelby (1957), and Bristow (1960) on solids with spherical pores, ellipsoidal inhomogeneities, and microcracks, respectively. For discussions of micromechanics issues in this broader context, see e.g. the reviews of Hashin (1983), Markov (2000), and Kachanov and Sevostianov (2005). In rock mechanics, problems of kind were first addressed by Walsh (1965 a,b). In the exploration geophysics community, similar results are often attributed to Hudson (1980) and Schoenberg (1980). Theoretical investigations into the effective elastic properties of cracked solids have been reviewed by Kachanov (1992,

1994) and, in the context of rock mechanics, in a tutorial by Grechka and Kachanov (2006c).

For the typically irregular microstructure of rocks, the main difficulty lies in finding a *quantitative characterization* of the microstructure, i.e., *identifying microstructural parameters in terms of which the effective elastic properties can be expressed*[1]. While this problem is far from fully solved, substantial progress has been made in this direction. For example, an important theoretical tool has been provided by the so-called comparison (or auxiliary) theorem of Hill (1963), which bounds the contribution from a pore to the overall compliance by those from inscribed and circumscribed pore shapes (see also Huet *et al.*, 1991). Often, however, theoretical analyses have to be supplemented by three-dimensional computational studies. It thus appeared timely at this stage to offer a critical assessment together with an identification of the remaining challenges.

Microstructural defects in the form of both cracks and pores are discussed in this review, but a key finding will be that *the dominant influence on the effective elastic properties of rocks comes from cracks, not pores.* This is demonstrated, for example, by the experimental results of Fortin *et al.* (2007) on isotropic compaction of a porous sandstone of 25% porosity (cf. Figure 3). In these experiments the porosity decreased by several percent. The elastic stiffnesses nevertheless decreased due to the development of new cracks. These cracks contributed to porosity in a negligible way (of the order of 0.1%), but were found to reduce the stiffnesses significantly. At sufficiently high pressure (of the order of 200 MPa), cracks close. We are interested here in the pressure range where they are open. In that pressure range, the presence or absence of cracks of negligible volume will account for a substantial difference in elastic properties.

This review focuses on *linear elastic* models that will be appropriate for a certain magnitude of the applied stresses. If the latter are compressive and sufficiently high, they may lead to substantial crack closure and, as a consequence, to a non-linear stress-strain relation. For linear elastic models to be applicable, compressive applied stresses will therefore have to remain sufficiently low in comparison with stress levels that cause noticeable crack

[1]In this chapter the term *microstructural parameter* is applied to macroscopic parameters that characterize a microstructure in some average sense, to be made precise in each case. Properly speaking, these are structure parameters that measure in some average sense one or more geometrical aspects of a given microstructure. In the simplest case, a parameter such as the porosity will provide only the void volume fraction, with no further structural information on the shape of the voids space. Porosity is therefore a parameter that may be expected to be of limited use for characterizing macroscopic properties, as is discussed further below.

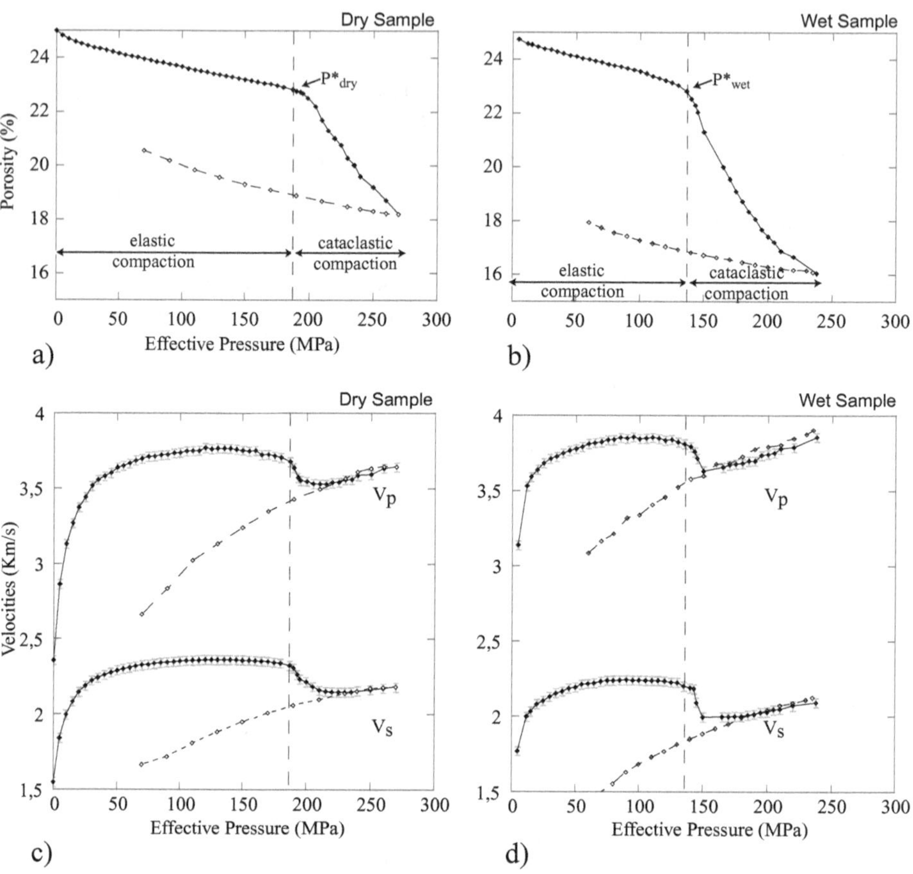

Figure 3. (a) Porosity drop due to grain crushing and pore collapse at P^* in a dry porous sandstone (Bleurswiller sandstone, initial porosity 25%, isotropic stress); (b) Porosity drop in the same sandstone, saturated; (c) V_P and V_S drop in the dry case; (d) V_P and V_S drop in the saturated case. Note that the P^* value is significantly lower in the saturated case (160 MPa) than in the dry one (220 MPa). Dotted curves correspond to unloading. Much lower velocities values in the unloading phase indicate a strong increase of crack apertures (from Fortin *et al.*, 2007).

closure effects. This is typically the case for ultrasonic elastic waves or seismic waves for which stress amplitudes are quite low.

Ultrasonic and seismic frequencies differ by six orders of magnitude and *frequency effects* are known to exist in rocks, due mainly to the presence of fluids. At ultrasonic frequencies (10^6 MHz), fluid flow cannot 'follow' the load so that the fluid pressure generally differs from pore to pore. This invalidates the fundamental *local equilibrium assumption* of poroelasticity theory (cf. Chapter 1) according to which the fluid pressure remains uniform, if time dependent, across a representative volume element (RVE). In this case, the elastic moduli are the 'unrelaxed' moduli and classical poroelasticity does not apply. At lower frequencies, fluid pressure has the time to equilibrate between pores on the scale of an RVE and poroelasticity will apply, the elastic moduli becoming the 'relaxed' moduli of poroelasticity. Relaxed moduli are defined for two different thermodynamic deformation conditions, viz. *drained* and *undrained* (cf. Chapter 1; see also Guéguen *et al.*, 2004, and Jaeger *et al.*, 2007). The critical frequency at which this local relaxation of pore pressure differences becomes effective varies in a broad range between 1 kHz and 1 MHz and depends, in particular, on the crack aspect ratio (Le Ravalec *et al.*, 1996). This implies that high-frequency laboratory data can only be compared to low-frequency field data, if frequency effects are properly accounted for (Le Ravalec and Guéguen, 1996a; Schubnel and Guéguen, 2003).

Although the rock matrix that forms the 'background' of the defects considered in the following will mostly be isotropic, the case of an anisotropic background is also briefly discussed. On the macroscale, anisotropy may result both from the microstructure (non-random orientations of cracks) and from the anisotropy of the matrix background. Some closed form results for this dual anisotropy are available for 2-D problems (Mauge and Kachanov, 1994). In the 3-D case, problems of this kind are much more difficult to deal with. There is only one simple situation for which an exact solution is known, i.e., the case of a transversely isotropic background rock containing cracks that run parallel to the symmetry plane (Laws, 1985; Sarout and Guéguen, 2008b). However, as has been shown by Ougier-Simonin *et al.* (2009), substituting an equivalent isotropic matrix for the real anisotropic matrix will often produce a good approximation.

In rocks the background of the defects consist typically of a polycrystalline matrix, which is therefore also inhomogeneous. The assumption of a homogeneous, anisotropic matrix will thus be rather artificial in most cases. But inhomogeneity of a random kind opens the possibility of statistical homogeneity and isotropy, and it is in the sense of a hypothetical equivalent of a statistically homogeneous, isotropic background that the assumed homogeneity and isotropy of the matrix may be interpreted in the following. A justification of this interpretation remains of course to be given in specific cases.

2 Quantitative Characterization of Microstructures: General Considerations

A quantitative characterization of a microstructure is contingent upon the identification of one or more microstructural parameters (such as defect concentration parameters) that determine the value of an effective elastic property *uniquely* by some functional relationship (see Kachanov and Sevostianov, 2005, for further detail)

$$\text{effective property} = f(\text{microstructural parameters}). \tag{1}$$

Such characterizations will in general *differ for different physical properties.* For example, the width of cracks and their interconnectedness are essential determinants of an effective permeability, but will affect elastic properties to a much lesser extent. Another example is the sharpness of crack tips or that of various corners at pore walls, which does affect the effective linear elastic properties only little, while it is of primary importance to fracture, i.e. to crack growth. Given these complexities, the identification of the argument(s) of the function (1) in general poses a challenging problem. There also arises the issue of overall anisotropy in this context, since the latter is determined by the tensorial rank and symmetry of the mentioned argument, as well as by the anisotropy of the background rock. The key qualification of a proper microstructural parameter is its ability to represent individual defects in accordance with their actual contribution to an effective property.

Microstructural parameters such as a volume fraction or–in the simplest case–a crack density are typically introduced within the framework of a non-interaction approximation (NIA). This is perhaps the only realistic approach, since interaction effects governed by the statistics of the mutual positions of inhomogeneities are obviously not accounted for by a volume fraction or crack density. One therefore has to be aware of a limitation of (1), which lies in the inconsistency of accounting for interactions via a certain choice of the function f of arguments that cannot capture the necessary detail.

2.1 Simple Microstructural Parameters and their Limitations

Appropriate microstructural parameters for characterizing effective elastic properties have been identified for the simplest defect shapes.

Porosity. Under certain conditions of vanishing interaction effects, to be discussed further below, the volume fraction (porosity) of N *spherical pores* of volume $V^{(m)}$ $(m = 1, \ldots, N)$ in a representative volume V, as given by

$$\phi = \frac{1}{V} \sum_{m=1}^{N} V^{(m)}, \tag{2}$$

will suffice (together with the matrix moduli) to characterize an effective elastic property. This use of the porosity can be somewhat extended beyond strictly spherical pores, in the following situations:

- Moderately non-spherical pores (for example, spheroids with aspect ratios[2] in the range 0.7 - 1.4). For a given volume fraction, these will produce approximately the same effect on the overall elasticity as spherical pores, provided the orientations of the non-sphericities are random (Kachanov *et al.*, 1994).
- A moderate surface roughness of the pore walls. That this factor is of little consequence follows from the above-mentioned comparison theorem of Hill (1963).
- All pores have the same shape and orientation. The effective elastic constants are then functions of ϕ and an additional shape parameter. This shape factor can be identified and the function (1) can be determined for ellipsoidal pores.

The use of a volume fraction as a sole parameter in (1) has its limitations, i.e.:

- It cannot cover cases in which the bulk response becomes anisotropic, as with non-spherical pores that exhibit preferred orientations. These will demand tensorial microstructural parameters.
- For mixtures of diverse pore shapes, shape parameters are needed, even in the case of an overall elastic isotropy (cf. the discussion of Section 3).

Crack density. When dealing with N *circular cracks* with radius $a^{(m)}$ $(m = 1, \ldots, N)$ in a representative volume V, the crack density

$$\rho = \frac{1}{V} \sum_{m=1}^{N} a^{(m)3} \tag{3}$$

[2] Note that *aspect ratios* are defined as 'width over length' in this chapter.

serves as microstructural parameter. In the 2-D case of rectilinear cracks of lengths $2a^{(m)}$, one has $\rho = 1/A \sum_m a^{(m)2}$, A being a representative area. Note that in (3) the contributions from individual cracks are proportional to the cube of their radius, which implies that small cracks can be ignored in presence of much larger ones unless they vastly outnumber them.

The crack density parameter is sometimes written as $N < a^3 >$. This form seems less desirable since it makes it less clear that small cracks can be ignored. Note also that the average $< a^3 >$ is sometimes erroneously replaced by $<a>^3$, which differs from $<a^3>$.

The parameter ρ was introduced by Bristow (1960) to materials science and by Walsh (1965a,b) in the context of geophysical applications; it was extended to the elliptic cracks by Budiansky and O'Connell (1976), who defined

$$\rho = \frac{2}{\pi V} \sum_{m=1}^{N} \left(\frac{A^2}{P} \right)^{(m)}, \tag{4}$$

where $A^{(m)}$ and $P^{(m)}$ denote the crack area and the perimeter of the mth crack. When characterizing effective elastic properties, an important observation is that a distribution of *multiple* elliptic cracks can be replaced, with good accuracy, by an equivalent distribution of circular cracks, provided the axis orientations of the ellipses are randomly distributed; the crack density parameter ρ of this equivalent distribution can then be explicitely expressed in terms of the ellipse axes (Kachanov, 1992, 1994). The possibility of such equivalent replacements for more general, 'irregular' crack shapes that are of interest in geophysical applications, is discussed further in Section 3.2.

For a unified coverage of non-random orientations, the scalar crack density ρ was obviously inadequate and was generalized for this purpose by Kachanov (1980), who introduced the second rank *crack density tensor*

$$\boldsymbol{\alpha} = \frac{1}{V} \sum_{m=1}^{N} a^{(m)3} \mathbf{n}^{(m)} \mathbf{n}^{(m)} \quad \text{or shorter} \quad \boldsymbol{\alpha} = \frac{1}{V} \sum_{m} (a^3 \mathbf{nn})^{(m)}, \tag{5}$$

where $\mathbf{n}^{(m)}$ is the unit normal to the mth crack and $\mathbf{n}^{(m)}\mathbf{n}^{(m)}$ - or $(\mathbf{nn})^{(m)}$ - is the dyadic product[3] with Cartesian components $(\mathbf{nn})_{ik}^{(m)} = n_i^{(m)} n_k^{(m)}$. In the 2-D case one has $\boldsymbol{\alpha} = 1/A \sum_m (a^2 \mathbf{nn})^{(m)}$. The linear invariant $\alpha_{kk} = \rho$ so that $\boldsymbol{\alpha}$ is indeed a natural tensorial generalization of ρ.

The fourth rank tensor

$$\boldsymbol{\beta} = \frac{1}{V} \sum_{m} (a^3 \mathbf{nnnn})^{(m)} \tag{6}$$

[3]The tensor character of the dyadic product of two vectors **a** and **b** is often emphasized by writing $\mathbf{a} \otimes \mathbf{b}$ for **ab**, using the tensor product operator $\otimes$.

was identified by Kachanov as a second crack density parameter. It plays a relatively minor role in dry rock or in the presence of highly compressible pore fluids in the sense that $\boldsymbol{\beta}$ enters in an effective elastic compliances with a relatively small factor, as will be discussed below. For liquid-saturated rocks, however, the contribution to the overall elastic properties arising from the parameter $\boldsymbol{\beta}$ may become important (see Section 4.2).

In 'dry' rock, for which only $\boldsymbol{\alpha}$ matters in the first approximation, the advantages of employing the crack density tensor $\boldsymbol{\alpha}$ as the sole microstructural parameter are the following:

- Any given orientation distribution of cracks is characterized by at most three parameters—the principal values of $\boldsymbol{\alpha}$. They reflect, in an integral way, all the details of the distribution that control the effective elasticity.
- Given that $\boldsymbol{\alpha}$ is a symmetric second rank tensor, the contribution to the overall elastic compliance from microcracks of any orientation distribution will exhibit an orthorhombic symmetry (orthotropy), the orthotropy axes being coaxial with the principal axes of $\boldsymbol{\alpha}$. This may not be intuitively obvious for it applies to cases where, *geometrically*, the distribution pattern may not display an orthorhombic symmetry. Moreover, this orthotropy will be of a special kind, being characterized by four independent elastic constants, each depending on five parameters, viz. the three principal values of $\boldsymbol{\alpha}$ and the two elastic constants of the isotropic background matrix (Kachanov, 1992).

This application of the crack density tensor $\boldsymbol{\alpha}$ and implications for the overall orthotropy can be somewhat extended to cracks that are not strictly circular and ideally thin, as follows:

- The tensor $\boldsymbol{\alpha}$, defined for cracks that are ideally thin, can also be used to characterize the effect of narrow, crack-like dry pores with aspect ratios up to about 0.1. An analysis shows that the contribution from such pores to the overall *linear* elastic compliance is almost the same as that from cracks. The porosity associated with these crack-like pores therefore becomes an irrelevant parameter when they are dry.
- For cracks with rough, corrugated surfaces the roughness factor can be ignored and the tensor $\boldsymbol{\alpha}$ be applied as long as the roughness does not generate contacts between crack faces. This follows from Hill's (1963) comparison theorem, which bounds the compliance of such a crack by those of an ideally thin crack and of a narrow pore, and from the above observation on narrow pores.

In more complex situations, such as when the cracks contain a pore fluid or are not flat, or when the background matrix is anisotropic, it may not be possible any more to characterize microstructural effects in terms of the a single crack density parameter $\boldsymbol{\alpha}$. These situations are discussed in Sections 4 and 5.

2.2 Microstructural Parameters for Pores and Cracks of Diverse Shapes and Orientations

Finding appropriate microstructural parameters for pores and cracks of diverse and 'irregular' shapes and preferred (non-random) orientations is not a trivial problem. Certain basic points therefore deserve to be discussed first, along with the progress made so far (see also Kachanov and Sevostianov (2005) for further detail).

The contribution of an individual defect to the effective elastic properties amounts to either a change in the overall compliance or in the overall stiffness of an RVE. In the asymptotic ('dilute') limit of small defect density the two formulations produce identical results. While this would remain true in an exact theory valid for any defect density, the same cannot be expected for approximate schemes. It is therefore of practical importance to know which of the two formulations will yield a larger value of defect density at which the 'dilute limit' remains sufficiently accurate. There appears to be no answer to this question for inhomogeneities in general. For cracks and pores, however, a compliance formulation is clearly preferable. Indeed, from a physical point of view, the natural choice is to treat cracks and pores as sources of an extra strain that can be expressed directly in terms of displacements at the defect boundaries. Summing over multiple defects then implies summing over their individual contributions to the overall compliance. This also implies that the effective compliances–not stiffnesses–are linear in the crack density. Direct computational studies for multiple cracks (Grechka and Kachanov, 2006a,b), discussed in the following, confirm that in a compliance formulation the non-interaction approximation (NIA) remains accurate up to much larger crack densities than it will in a stiffness formulation.

If an isolated cavity is placed in a solid under stress, the displacement of the material points on its surface S will contribute an extra strain to the overall strain per unit reference volume V, which is given by the surface integral

$$\Delta\boldsymbol{\varepsilon} = \frac{1}{2V}\int_S (\mathbf{un} + \mathbf{nu})dS \quad \text{or} \quad \Delta\varepsilon_{ij} = \frac{1}{2V}\int_S (u_i n_j + n_i u_j)dS, \tag{7}$$

where $\mathbf{u}$ is the displacement vector generated by applied stress $\boldsymbol{\sigma}$. For a crack, (7) reduces to

$$\Delta \varepsilon = \frac{1}{2V} \int_S ([\mathbf{u}]\mathbf{n} + \mathbf{n}[\mathbf{u}]) dS, \tag{8}$$

where S now stands for the oriented (by the unit vector $\mathbf{n}$) discontinuity surface that defines the crack (its surface area, which is also denoted by S, is that of one crack face) and $[\mathbf{u}] = \mathbf{u}^+ - \mathbf{u}^-$ is the displacement discontinuity vector along S. For a *flat (planar)* crack $\mathbf{n}$ is constant and

$$\Delta \varepsilon = (\mathbf{bn} + \mathbf{nb}) S/2V, \tag{9}$$

where $\mathbf{b} = < \mathbf{u}^+ - \mathbf{u}^- >$ is the average displacement discontinuity vector over S.

If there are N flat cracks in V, the extra strain due to these cracks becomes $\Delta \varepsilon = \frac{1}{2V} \sum_{(m)=1}^{N} (\mathbf{bn} + \mathbf{nb})^{(m)} S^{(m)}$ and this sum may be replaced by an appropriate integral over orientations, if computationally convenient.

Historical remark. The above expressions for the extra strain due to the presence of a pore or crack are an immediate consequence of a footnote remark of Hill (1963). In the explicit form, they started to appear in 1970's (see, for example, Vavakin and Salganik, 1975).

The overall strain of an elastic solid can be represented as the sum of an average matrix strain and the 'extra strain' (7) due to the presence of defects, writing

$$\varepsilon_{ij} = S^0_{ijkl} \sigma_{kl} + \Delta \varepsilon_{ij}, \tag{10}$$

where S^0_{ijkl} are the components of the matrix compliance and σ_{kl} are those of the applied stress.

The fundamental quantity to be found is then the *extra compliance* ΔS_{ijkl} resulting from the presence of pores and/or cracks, such that

$$\Delta \varepsilon_{ij} = \Delta S_{ijkl} \sigma_{kl} \tag{11}$$

and consequently

$$\varepsilon_{ij} = S_{ijkl} \sigma_{kl}, \tag{12}$$

where

$$S_{ijkl} = S^0_{ijkl} + \Delta S_{ijkl} \tag{13}$$

is now the *effective compliance* of the material.

2.3 The Non-interaction Approximation. Its Forms and its Role

The task of finding the extra compliance due to the presence of defects is greatly simplified within the framework of a *non-interaction approximation*

(NIA). Such an approximation places individual defects into the remotely applied stress field that is assumed to remain unperturbed by neighboring defects. The extra compliance resulting from all defects is then simply given by the sum of the contributions from all m individual defects per unit volume, enabling one to write

$$\Delta S_{ijkl} = \sum_m \Delta S_{ijkl}^{(m)}, \tag{14}$$

where the summation may be replaced by an integration over orientations, if computationally convenient.

In the following, we consider methods of finding concrete expressions for (14) for various classes of defects. In particular, we shall examine the possibility of expressing (14) in terms of a symmetric 2nd rank tensor. Simplifications of this kind are best identified in terms of the elastic potential $f(\boldsymbol{\sigma}) = \frac{1}{2}\sigma_{ij}S_{ijkl}\sigma_{kl}$, representing it as a sum

$$f = f_0 + \Delta f \tag{15}$$

with

$$f_0 = \frac{1}{2}\boldsymbol{\sigma} : \mathbf{S^0} : \boldsymbol{\sigma} = \frac{1}{2}\sigma_{ij}S_{ijkl}^0\sigma_{kl} = \frac{1+\nu_0}{2E_0}\sigma_{ij}\sigma_{ij} - \frac{\nu_0}{2E_0}\sigma_{kk}^2 \tag{16}$$

$$\Delta f = \frac{1}{2}\boldsymbol{\sigma} : \Delta\mathbf{S} : \boldsymbol{\sigma} = \frac{1}{2}\sigma_{ij}\Delta S_{ijkl}\sigma_{kl} \tag{17}$$

in the case of an isotropic rock matrix, with Young's modulus E_0 and Poisson ratio ν_0.

The formulation in terms of potential Δf, being equivalent to the one in compliances, has the advantage that it enforces the necessary symmetrization of (14), with respect to the first and second pairs of indices. As seen in the text to follow, this aids in identifying possible simplifications in the proper microstructural parameters via the structure of Δf.

Remark. In the case of non-negligible defect interactions, the tensor $\Delta \boldsymbol{S}$ of extra compliances will depend on the positions of defects. Various approximate approaches have addressed the effect of interactions, among which self-consistent (Hill, 1965; Budiansky and O'Connell, 1976), differential (Salganik, 1975; Bruner, 1976; Hashin, 1983; Le Ravalec and Guéguen, 1996), and Mori-Tanaka schemes (Mori and Tanaka, 1973; Benveniste, 1986) account for interactions by treating the effective elastic constants as functions of the parameters defined in the NIA; hence none of these schemes accounts for the statistics of defects positions. While such an approach may in fact be the only practical one, one should also be aware of its limitations (see Section 3.3 and Kachanov and Sevostianov (2005) for a more detailed discussion). We therefore mention a more advanced and mathematically more complex method of effective field due to Kanaun and Levin (see their book of 2008) that takes these position statistics into account.

In certain cases, such as in the presence of stiff inclusions, a summation over the contributions from individual defects to an overall *stiffness* will be more appropriate. In the NIA one then has

$$\Delta\sigma_{ij} = \Delta C_{ijkl}\varepsilon_{kl}, \tag{18}$$

where the *extra stiffness* is the sum

$$\Delta C_{ijkl} = \sum_m \Delta C_{ijkl}^{(m)} \tag{19}$$

over the contributions from all m inhomogeneities per unit volume.

The NIA (formulated in compliances) yields effective compliances S_{ijkl} that are linear in the defect concentration. For any S_{ijkl} component

$$S = S_0 + Lx, \tag{20}$$

where L is certain constant (specific for this component) and x is the appropriate defect density parameter. This implies that the effective stiffnesses C_{ijkl} (for example, Young's or shear moduli) have the form

$$C = \frac{C_0}{1 + Mx}, \tag{21}$$

where M is a constant.

A dual formulation in terms of the extra stiffness tensor $\Delta\boldsymbol{C}$ is sometimes used, where effective *stiffnesses* are linear functions of the defect concentration (see, for example, Hudson, 1980), implying that compliances, and not stiffnesses, have the form (21). As discussed above, this could be an appropriate choice if the inclusions are stiffer than the matrix. This formulation yields stiffnesses in the form

$$C = C_0(1 - Mx). \tag{22}$$

In the asymptotic limit of small defect density this is obviously equivalent to (21). An important difference between the two formulations is that for cracks, the linearization (22) drastically narrows the range of defect densities in which the NIA results remain accurate, whereas the formulation in compliances, which leads to (21), remains accurate up to substantial crack densities (Grechka and Kachanov, 2006a). We observe that the version (22) predicts zero stiffnesses at some density value $x_c = M^{-1}$ and negative stiffnesses beyond. For cracks, x_c corresponds to crack densities that are quite low by geophysical standards; a similar conclusion would hold for pores of various shapes. Thus, for cracks and pores, a formulation in compliances, (15) & (20), will be the preferred choice.

Remark. The critical value x_c is sometimes interpreted as marking a percolation threshold (Mukerji *et al.*, 1995). This interpretation is incorrect, since crack intersections are not accounted for by models of the present kind, as has been discussed by Guéguen *et al.* (1997); it is simply a consequence of linearization. The same remark would also apply to x_c predicted by the self-consistent scheme, where it emerges as a result of a particular scheme of calculating the effective constants.

In effective medium theories, the NIA plays a central role for the following reasons:

- It is rigorously correct in the limit of small defect density.
- It constitutes the basic building block for various approximate effective media schemes (self-consistent, differential, Mori-Tanaka), that place non-interacting defects into some sort of effective environment (effective matrix or effective stress).
- Computational studies show that, for *cracks*, predictions based on a NIA yield satisfactory accuracy at moderate crack densities when the *local* effects of crack interactions become strong. This is true, if the mutual positions of cracks are uncorrelated (random) so that, for example, periodic arrangements are excluded.

3 Dry Rock with Cracks

This section focuses on the determination of the effective elastic compliances of rocks with cracks and pores of various shapes (Figure 4), using a non-interaction approximation (NIA). It is restricted to the case of an isotropic rock matrix, except for Section 3.4 where the effect of matrix anisotropy is briefly discussed. We shall first deal with cracks, paying attention also to irregular crack shapes, and thereafter with pores of diverse shapes. We consider only dry rock in this section, postponing a discussion of fluid-filled pores and cracks until Section 4.

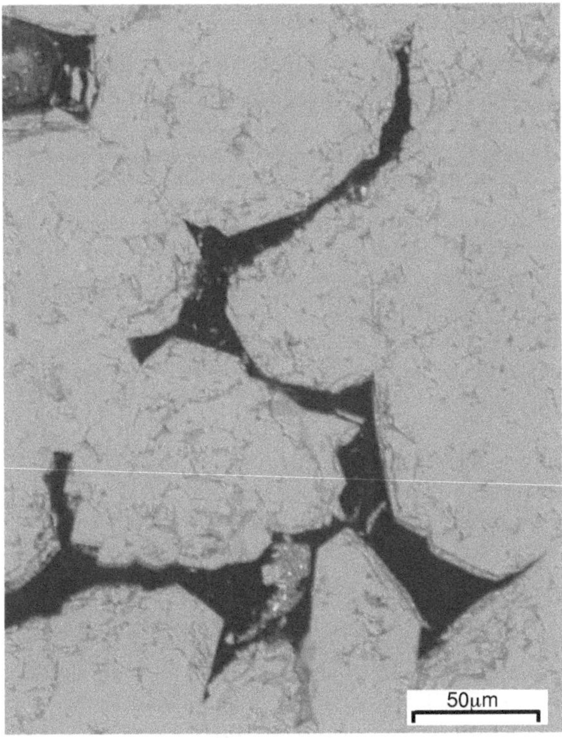

Figure 4. SEM image of cracks and pores in deformed Fontainebleau sandstone at 17% porosity. Quartz grains and SiO_2 cement. Note densely cracked grain boundaries. Crack apertures are strongly enhanced due to unloading, in agreement with the data of Figure 3. (Courtesy J. Fortin)

Remark. A related problem, which has received much attention in the rock mechanics literature, is that of determining the effective stiffness of a rock mass containing rough joint faces pressed against one another. We note that this situation is distinctly different from problems that involve traction-free cracks in that the effective stiffness will be controlled not by crack density, but by the statistics of contacts (Kachanov et al, 2010).

3.1 Cracks of Circular and Elliptic Shape. Crack-Induced Anisotropy

For planar cracks of circular or elliptical shape, results based on a NIA assume a simple form for any crack orientation distribution. Computational studies of the accuracy of a NIA, i.e. studies that seek to determine the crack density at which a NIA becomes inaccurate, are discussed in Section 3.3.

The extra compliance generated by a single planar crack. The strain generated by a flat crack in response to a remotely applied stress $\boldsymbol{\sigma}$ is determined, according to (9), by the average displacement discontinuity $\mathbf{b}$. A symmetric second rank crack compliance tensor $\mathbf{B}$ can be introduced that relates $\mathbf{b}$ linearly to the uniform resolved traction $\mathbf{t}_{(\mathbf{n})} = \mathbf{n} \cdot \boldsymbol{\sigma}$ of the remote stress at the crack site, according to

$$\mathbf{b} = \mathbf{B} \cdot \mathbf{t}_{(\mathbf{n})} \tag{23}$$

The tensor $\mathbf{B}$ was first introduced by Schoenberg (1980) and denoted by $\mathbf{Z}$. The challenge lays however in determining $\mathbf{B}$ in terms of crack geometry. For the elliptic cracks, in particular, $\mathbf{B}$ was given by Kachanov (1992) based on Eshelby results. In Section 3.2 further results are given for cracks of various irregular shapes, such as cracks with partial contacts between crack faces.

In coordinate system $(\mathbf{n}, \mathbf{t})$, where $\mathbf{t}$ is tangential to the crack, let B_{nn} and B_{tt} denote the normal and a shear compliance (depending on the direction of $\mathbf{t}$) of a crack. These give the normal and shear components of $\mathbf{b}$ produced by uniform unit tractions applied in the two directions. The off-diagonal components $B_{nt} = B_{tn}$ then characterize the coupling of the normal and shear modes. If the matrix is isotropic, $B_{nt} = 0$.

Since $\mathbf{B}$ is a symmetric second rank tensor, there exist three mutually orthogonal principal directions of the crack compliance, such that the application of a uniform traction in one of these directions does not generate components of $\mathbf{b}$ in the other two directions. In the case of the isotropic matrix, $\mathbf{n}$ is one principal direction, the other two being given by unit vectors (eigenvetors) $\mathbf{s}$ and $\mathbf{t}$ that lie in the crack plane. We thus have the representation

$$\mathbf{B} = B_{nn}\mathbf{nn} + B_{ss}\mathbf{ss} + B_{tt}\mathbf{tt} \tag{24}$$

Following Sayers and Kachanov (1995), we now introduce the arithmetic mean of the principal shear compliances

$$B_T = \frac{B_{ss} + B_{tt}}{2} \tag{25}$$

For the elastically axisymmetric crack shapes (this includes not only a circle, but certain shapes that have lesser symmetries, for example, the symmetry of any equilateral polygon), $B_{tt} = B_{ss} \equiv B_T$, independent of the in-plane direction, so that

$$\mathbf{B} = B_N \mathbf{nn} + B_T(\mathbf{I} - \mathbf{nn}) \tag{26}$$

where $B_N \equiv B_{nn}$. The following important question now arises: Under which conditions are the compliances B_N and B_T equal (so that $\mathbf{B}$ is isotropic), at least approximately? For if $B_N \approx B_T$, then any crack orientation distribution may approximately be characterized by the symmetric 2nd rank crack density tensor $\boldsymbol{\alpha}$, implying overall elastic orthotropy as a far-reaching consequence. Moreover, this crack-induced orthotropy will be of a special type, being characterized by a reduced number of elastic constants. These conclusions remain valid even if $B_N \approx B_T$ holds only on average for a set of cracks, as is discussed further below.

The compliances B_N and B_T are exactly equal in the 2-D case of a rectilinear crack of length $2a$. In this case one has, for a plane state of stress,

$$\mathbf{B} = (\pi a/E_0)\mathbf{I}. \tag{27}$$

For a *dry circular crack* of radius a in three dimensions (see, for example, Rice, 1979)

$$B_T = \frac{32(1-\nu_0^2)a}{3\pi E_0(2-\nu_0)}, \qquad \frac{B_N}{B_T} = 1 - \nu_0/2. \tag{28}$$

The values of B_N and B_T are relatively close (the factor $\nu_0/2$ is of the order of 0.1 for $\nu_0 = 0.25$, a typical value for rocks) and the deviation of $\mathbf{B}$ from isotropy is relatively small.

For *elliptical* cracks the tensor $\mathbf{B}$ was derived by Kachanov (1992) on the basis of Eshelby's results for ellipsoidal inhomogeneities[4] as specialized for cracks by Budiansky and O'Connell (1976). Taking $\mathbf{s}$ and $\mathbf{t}$ as unit vectors along the major and minor axes of length $2a$ and $2b$, respectively, the principal shear compliances B_{ss} and B_{tt} in (24) are expressed in terms of complete elliptic integrals of the argument $k = \sqrt{1-(b/a)^2}$. An important finding is that the average shear compliance B_T and B_N are about as close as they are for circular cracks. This allows one to represent *multiple* elliptical cracks with random eccentricities by an equivalent distribution of circular cracks with radii explicitly expressed in terms of the ellipses' axis lengths.

In general, it follows from (9) and (23) that the contribution of a single flat crack of arbitrary shape to the overall strain, when expressed in terms of the tensor $\mathbf{B}$, becomes

$$\Delta\varepsilon = \frac{S}{V}\mathbf{nBn} : \boldsymbol{\sigma}, \tag{29}$$

[4]See Chapter 2.

so that in accord with (11) the extra compliance resulting from the presence of the crack is given by

$$\Delta \mathbf{S} = \frac{S}{V}\mathbf{nBn} \quad \text{or} \quad \Delta S_{ijkl} = \frac{S}{V} n_i B_{jk} n_l, \tag{30}$$

symmetry with respect to the exchanges $ij \leftrightarrow kl, i \leftrightarrow j, k \leftrightarrow l$ being imposed.

Crack-induced anisotropy. Orthotropy and transverse isotropy. We now consider the general case of non-random crack orientations and crack-induced anisotropy and examine ways in which it may be simplified for cracks. We shall find that the possibility of a simplification will depend on the closeness of the normal and shear crack compliances B_N and B_T.

For multiple flat cracks, the NIA applied to (30) yields the extra compliance

$$\Delta \mathbf{S} = \frac{1}{V}\sum_m (S\mathbf{nBn})^{(m)} \tag{31}$$

for flat cracks of arbitrary shape.

Assume first that **B**-tensor is isotropic, with $\mathbf{B} = k\mathbf{I}$. The extra compliance

$$\Delta \mathbf{S} = \frac{1}{V}\sum_m (kS\mathbf{nIn})^{(m)} \tag{32}$$

is thus expressible in terms of a symmetric 2nd rank tensor as the only relevant crack density parameter, a simplification of (31) that hinges upon the equality of B_N and B_T.

Next consider a circular crack of radius a for which B_N and B_T differ, but are relatively close. Writing (26) as $\mathbf{B} = B_T[\mathbf{I} - (\nu_0/2)\mathbf{nn}]$ for this case and using (28), one finds (Kachanov, 1980)

$$\Delta \mathbf{S} = \frac{32(1-\nu_0^2)}{3(2-\nu_0)E_0}\left[\frac{a^3}{V}(\mathbf{nIn} - \frac{\nu_0}{2}\mathbf{nnnn})\right] \tag{33}$$

or

$$\Delta S_{ijkl} = \frac{32(1-\nu_0^2)}{3(2-\nu_0)E_0}\left[\frac{1}{4}(\delta_{ik}\alpha_{jl} + \delta_{il}\alpha_{jk} + \delta_{jk}\alpha_{il} + \delta_{jl}\alpha_{ik}) - \frac{\nu_0}{2}\beta_{ijkl}\right]. \tag{34}$$

We see that, in addition to the second rank crack density tensor $\boldsymbol{\alpha} = \frac{1}{V}\sum_m (a^3\mathbf{nn})^{(m)}$, the fourth rank tensor $\boldsymbol{\beta} = \frac{1}{V}\sum_m (a^3\mathbf{nnnn})^{(m)}$ now enters the extra compliance as a second crack density parameter.

Neglecting the $\boldsymbol{\beta}$-term and retaining $\boldsymbol{\alpha}$ as the sole crack density parameter constitutes an acceptable approximation in most cases. That this statement extends to flat cracks of *irregular* shape provided the irregularities are of a random kind is implied indirectly by wave speed data (Sayers and Kachanov, 1995) and confirmed by direct computations (Grechka and Kachanov, 2006b). The crack-induced anisotropy is of a particularly simple form in this case. Here we outline the basic findings, referring for details to Kachanov's work of 1980 or his reviews of 1992 and 1993.

Since $\boldsymbol{\alpha}$ is a symmetric second-rank tensor, it is characterized by the three principal values $\alpha_1, \alpha_2, \alpha_3$. In the framework of the present approximation any orientation distribution of cracks is therefore equivalent to three mutually perpendicular families of parallel cracks with respective partial scalar densities $\rho_1 = \alpha_1, \rho_2 = \alpha_2, \rho_3 = \alpha_3$. This implies that a cracked (dry) solid displays an (approximately) orthotropic (orthorhombic) symmetry–a somewhat counterintuitive statement; for it applies to *geometrical* crack patterns that may not possess this symmetry (e.g., two crack families of different density, inclined at an arbitrary angle to one another). Moreover, this orthotropy is found to be of a simplified 'elliptic' type, in which the number of independent constants is reduced substantially, from nine to only four. Accordingly, when the $\boldsymbol{\beta}$-term is neglected, the extra compliances (33)–referred to the principal axes of $\boldsymbol{\alpha}$–assume the form

$$\left.\begin{aligned}
\Delta S_{1111} &= h\alpha_1 \\
\Delta S_{2222} &= h\alpha_2 \\
\Delta S_{3333} &= h\alpha_3 \\
\Delta S_{1212} &= \tfrac{h}{4}(\alpha_1+\alpha_2) &&= \tfrac{1}{4}(\Delta S_{1111}+\Delta S_{2222}) \\
\Delta S_{2323} &= \tfrac{h}{4}(\alpha_2+\alpha_3) &&= \tfrac{1}{4}(\Delta S_{2222}+\Delta S_{3333}) \\
\Delta S_{3131} &= \tfrac{h}{4}(\alpha_3+\alpha_1) &&= \tfrac{1}{4}(\Delta S_{3333}+\Delta S_{1111})
\end{aligned}\right\} \quad (35)$$

with $h = 32(1-\nu_0^2)/3E_0(2-\nu_0)$. Thus, only three of the components ΔS_{ijkl} are found to be independent. In the same coordinate system the overall compliances are given by

$$\left.\begin{aligned}
S_{1111} &= \tfrac{1}{E_0} + h\alpha_1 \\
S_{2222} &= \tfrac{1}{E_0} + h\alpha_2 \\
S_{3333} &= \tfrac{1}{E_0} + h\alpha_3 \\
S_{1122} &= S_{2233} = S_{3311} = -\tfrac{\nu_0}{E_0} \\
S_{1212} &= \tfrac{1+\nu_0}{2E_0} + \tfrac{h}{4}(\alpha_1+\alpha_2) &&= \tfrac{1}{4}(S_{1111}+S_{2222}-2S_{1122}) \\
S_{2323} &= \tfrac{1+\nu_0}{2E_0} + \tfrac{h}{4}(\alpha_2+\alpha_3) &&= \tfrac{1}{4}(S_{2222}+S_{3333}-2S_{2233}) \\
S_{3131} &= \tfrac{1+\nu_0}{2E_0} + \tfrac{h}{4}(\alpha_3+\alpha_1) &&= \tfrac{1}{4}(S_{3333}+S_{1111}-2S_{3311})
\end{aligned}\right\} \quad (36)$$

Of these nine elastic constants, only four are independent; they depend on the five parameters $(\alpha_1, \alpha_2, \alpha_3, \nu_0, E_0)$.

Remark. The three components $\alpha_1, \alpha_2, \alpha_3$, represent the information on crack distributions that can be extracted from the effective elastic constants and therefore from wave velocity data; no further information of the 'details' of crack distributions can be extracted.

Of particular interest for applications is the case of *transverse isotropy* (TI) or hexagonal symmetry. If we choose x_3 as the axis of symmetry, this covers several orientation distributions of cracks for which $\alpha_1 = \alpha_2$, for example,

- Approximately parallel cracks. This case is relevant, for example, for foliated rocks where cracks are almost exactly parallel.
- Crack normal orientations that are randomly distributed in the x_1, x_2 plane.

Note that both cases may involve random *orientation scatter*, i.e., scatter without directional bias.

The case of *randomly oriented* cracks (isotropy) is the simplest one. While not expected if crack orientations are stress controlled, as in typical tectonic environments, it may nevertheless be a realistic in the case of thermal cracking (Le Ravalec and Guéguen, 1994), or when compaction of a porous rock occurs in an isotropic stress field (Fortin *et al.*, 2007). In this case, the effective bulk K and shear G moduli involve the scalar crack density ρ as a single parameter and are given by

$$\frac{K_0}{K} = 1 + \frac{16}{9}\frac{(1-\nu_0^2)}{(1-2\nu_0)}\rho \quad \text{and} \quad \frac{G_0}{G} = 1 + \frac{16}{9}\frac{(1-\nu_0/5)(1-\nu_0)}{(1-\nu_0/2)}\rho. \tag{37}$$

3.2 Cracks of Irregular Shape

The scalar crack density parameter ρ and its tensorial generalization $\boldsymbol{\alpha}$ are defined for cracks of *circular* shape, with an extension to elliptic cracks due to Budiansky and O'Connell (1976). However, cracks that occur in rocks usually have an irregular shape, both at a micro- and macroscale. In rock mechanics this is often ignored and the conventional crack density parameter ρ is used. This implies treating the crack density as *adjustable parameter*. The link to the microstructure is thereby lost and the results become much less valuable.

Crack shape is a factor at least as important as crack interactions. Here we review the progress that has been made in an effort to quantify the effect of typical crack 'irregularity factors' on the effective properties to the microstructure.

Planar cracks of irregular shape It is clear that a *single* irregularly shaped crack cannot be replaced, in its contribution to the overall compliance, by an equivalent circular or even elliptical crack; for the three principal crack compliances cannot be matched by an appropriate choice of two axes of the ellipse. Two important questions, which may therefore be asked, are the following:

- Can a *distribution* of irregularly shaped cracks be replaced by an equivalent distribution of the circular cracks, given that their irregularities are of a random kind?
- If the answer to this question is in the affirmative, what is the 'effective density' of the equivalent distribution of the circular cracks?

We recall that, since $\mathbf{B} = B_N\mathbf{nn} + B_{ss}\mathbf{ss} + B_{tt}\mathbf{tt}$ is a symmetric second-rank tensor, any flat crack has three mutually perpendicular crack compliance directions and in the case of an isotropic host material $\mathbf{n}$ is the crack normal while $\mathbf{s}$ and $\mathbf{t}$ lie in the crack plane. On account of the identity $\mathbf{I} = \mathbf{nn} + \mathbf{ss} + \mathbf{tt}$ this implies

$$\mathbf{B} = B_T\mathbf{I} + [B_N - B_T]\mathbf{nn} + \frac{B_{ss} - B_{tt}}{2}(\mathbf{ss} - \mathbf{tt}), \tag{38}$$

where $B_T = (B_{ss} + B_{tt})/2$ is the average in-plane shear compliance.

The extra elastic potential (17) stored under stress in the presence of cracks now assumes the following form (Kachanov, 1992)

$$\begin{aligned}\Delta f &= \frac{1}{2V}(\boldsymbol{\sigma}\cdot\boldsymbol{\sigma}) : \sum_m (SB_T\mathbf{nn})^{(m)} \\ &+ \frac{1}{2V}\,\boldsymbol{\sigma} : \sum_m [S(B_N - B_T)\mathbf{nnnn}]^{(m)} : \boldsymbol{\sigma} \\ &+ \frac{1}{2V}\sum_m [S(B_{ss} - B_{tt})(\sigma_{ns}^2 - \sigma_{nt}^2)]^{(m)}.\end{aligned} \tag{39}$$

The first term in this expression contains a symmetric second-rank tensor as microstructural parameter of the $\boldsymbol{\alpha}$ type, the second term a fourth-rank tensorial parameter of the $\boldsymbol{\beta}$ type. The two terms are similar in structure to corresponding terms for circular cracks. The third term involves the differences $(\sigma_{ns}^2 - \sigma_{nt}^2)$ between squares of the remote shear stresses, resolved onto the crack planes in the principal directions $\mathbf{s}$ and $\mathbf{t}$ of the crack compliance tensor $\mathbf{B}$. If the crack shape irregularities are of a random kind, (uncorrelated with crack areas and orientations) the third term vanishes. The structure of Δf then resembles that of circular cracks, and complete agreement exists, if the ratio of the second term to the first one is as small as it is for circular cracks. Direct computational studies of Grechka *et al.*

(2006) confirm that this is indeed the case so that the answer to the first of the above questions is affirmative: when averaged over in-plane directions, the shear crack compliance is close to the normal compliance and typically as close as B_T and B_N for a circular crack.

For flat cracks with shape irregularities of a random kind, the following result for dry circular cracks therefore also holds approximately with sufficient accuracy: The $\boldsymbol{\alpha}$ tensor suffices to characterize the microstructure and orthotropy (with a reduced number of independent constants) of the bulk rock. This was indirectly verified by examining wavespeed patterns in microcracked rocks (Sayers and Kachanov, 1995) and further validated by the above-mentioned computational studies of irregularly shaped cracks (Grechka and Kachanov, 2006c).

As to the second of the above questions, the density of an equivalent distribution of circular cracks can be explicitely given for elliptical cracks in terms of ellipse axes. This can also be done for circular cracks with concentric 'islands' of contact between the faces, an important case that is discussed separately in section 3.2.3. A number of semi-rigorous estimates of B_N for several classes of irregular shapes were given by Sevostianov and Kachanov (2002); along with the approximation $B_N \approx B_T$ these estimates allow one to determine an equivalent density of circular cracks for these classes of crack shape.

Hill's bounds

When applied to cracks, this general theoretical tool bounds the contribution of an irregularly shaped crack to the total compliance by those of an inscribed and a circumscribed 'comparison shape'. Its usefulness is determined by the possibility to tightly bound a given shape by comparison shapes with known contributions to the compliance (cf. Fig. 5). The library of such comparison shapes is limited to elliptical cracks, annular cracks and certain classes of irregular crack shapes for which the above mentioned semi-rigorous estimates are available. Hill's theorem nevertheless provides certain useful insights, in particular:

- One shape factor that has been discussed in literature is the sharpness (or bluntness) of a crack tip (Mavko and Nur, 1978). However, the different geometry of sharp and blunt crack tips only affects the *non-linear* response under compression. As far as the linear elastic response (tangent stiffness) and wavespeeds are concerned, this factor is unimportant, since a crack with a sharp tip can be tightly bounded by two cracks with blunted tips and vice versa.
- The surface roughness or 'corrugation' of crack faces is not an important factor. This follows from the possibility of tightly bounding a

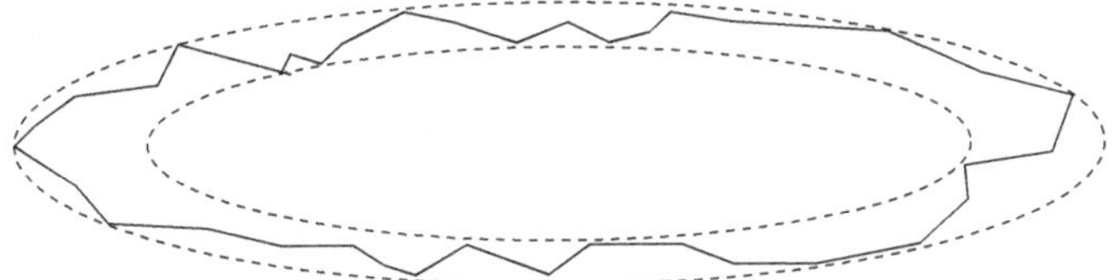

Figure 5. Irregularly shaped crack bounded by inscribed and circumscribed ellipses.

corrugated crack by an ideally thin crack and a strongly oblate pore and from the fact that the compliance of the latter comes close to that of the former. This assumes, of course, that no contacts between opposing crack faces are formed or deactivated and that no substantial flattening of existing contacts develops during loading, for then a linear elastic crack model would no longer apply.

Islands of contact between crack faces

Islands of contact or welded areas on opposing crack faces, even in the form of small asperity contacts, have a strong stiffening effect and consequently tend to reduce an equivalent density of circular cracks substantially. The normal compliance B_N of a circular crack with a contact area in the form of a concentric island is given by (Sevostianov and Kachanov, 2001)

$$B_N = \frac{2a(1-\nu_0^2)}{E_0} \frac{\varphi(\lambda)}{\lambda(2-\lambda)}. \tag{40}$$

Here $(a-c)$ and a are, respectively, the internal and external radii of the annular crack, $\lambda = c/a$ and

$$\varphi(\lambda) = -\lambda^3 \left(\frac{2\sqrt{2}}{\pi} - 1\right)^2 \ln(1-\lambda) + \frac{\lambda^2(2-\lambda)}{2} + 2\lambda^3 \left(\frac{2\sqrt{2}}{\pi} - 1\right) \tag{41}$$

The presence of a small island - implying a small departure of λ from 1 - substantially reduces the compliance. In the limit $\lambda \longrightarrow 1$ (vanishlingly small island), B_N should approach $16a(1-\nu_0^2)/3\pi E_0$, its value for the circular crack of radius a. Relation (40) does not yield this limiting value; moreover, the logarithmic term tends to infinity in this limit. This is due to the fact that the derivation of equation (40) involved certain numerical approximations. However, the approach of the logarithmic term to infinity is extremely slow; even at $1 - \lambda = 10^{-20}$ B_N still remains below its value for a circular crack. Hence, the results of (40) can be viewed as a good approximation even at very small 'islands'. Since the effect of a crack is proportional to its size cubed, *a very small island reduces the normal compliance roughly by a factor 2-3.* This implies that the existence of islands require a substantial adjustment of "effective" crack density. This adjustment will be approximately the same for the shear crack compliance, as indicated by the computational studies of Grechka *et al.* (2006). Therefore, characterization of multiple cracks of diverse orientations containing islands by crack density tensor $\boldsymbol{\alpha}$ can be retained.

Non-flat cracks Two main classes of non-flat geometries that are relevant to geophysical applications, viz., *intersecting flat cracks* and *wavy cracks*, have each formed the subject of three-dimensional computational studies by Grechka and Kachanov (2006a) and by Mear *et al.* (2007), respectively, with the following findings.

Intersections of flat cracks (that do not dissect the material into separate blocks) can be ignored in their impact on effective elastic properties, since these are little affected by the local stress field near crack intersection lines.

Remark. This is in contrast with the obvious importance of intersections to *transport properties* that are controlled by crack connectivity (Madden, 1983; Guéguen and Dienes, 1989).

For *wavy cracks*, small-to-moderate waviness can be ignored. A more pronounced waviness cannot be ignored, however, and such cracks can in general not be replaced by equivalent flat cracks. Whenever the crack geometries do not involve multiple waves with a high amplitude/wavelength ratio, models based on an equivalent distribution of flat cracks may still be possible. This includes, for example, cap-like geometries, single waves or multiple waves with moderate amplitude/wavelength ratios.

3.3 Crack Interactions

The effect of crack interactions on the overall elasticity has been discussed, mostly, in the general context of mechanics and materials science, for a number of years. The usual assumption has been that a non-interaction approximation loses accuracy at certain crack densities and a number of approximate schemes have therefore been proposed to account for interaction effects due to cracks and other inhomogeneities. Among these, the best known are the *differential scheme* (Vavakin and Salganik, 1975; Hashin,

1988), the *self-consistent scheme* (Budiansky and O'Connell, 1976), and the *Mori-Tanaka scheme* (Mori and Tanaka, 1973) which was clarified and specialized to cracked materials by Benveniste (1986). All of these schemes involve placing *non-interacting* cracks into certain 'effective' surroundings or environment, formed by an effective matrix (softened by cracks) in the first two schemes, and an effective stress field in the last scheme. Thus, in each of the mentioned schemes, the NIA forms the basic building block.

These three schemes produce substantially different predictions. The first two predict a softening effect of interactions; quite strong in the self-consistent scheme (which predicts zero stiffnesses at a certain crack density), this effect tends to be milder in the differential scheme. We note that these two schemes can only be set up either for random crack orientations (leading to overall isotropy) or for strictly parallel cracks, since they require solutions for one crack in the effective matrix. If the latter is anisotropic (due to intrinsic anisotropy or to non-random crack orientations), such solutions are available, in 3-D, only for crack running parallel to the isotropy plane in a transversely isotropic material.

The Mori-Tanaka scheme places cracks into an averaged solid matrix stress field. It has the advantage of being very simple and easily applicable to anisotropic cases of non-random defects orientations: since pores increase all the components of the mentioned average stress by the factor of $(1-\phi)^{-1}$, we have $\Delta S_{ijkl} = (1-\phi)^{-1}\Delta S_{ijkl}^{NIA}$. In the case of very thin cracks, porosity is negligible and this scheme coincides with the NIA, which translates into the statement that the competing interaction effects–stress shielding and stress amplification–cancel each other. Note that cracks with aspect ratios that are smaller than 0.1 but non-negligible will have compliance contributions close to the ones of ideally thin cracks. For them, however, the Mori-Tanaka scheme diverges from the NIA, in having the factor $(1-\phi)^{-1}$ where ϕ is due to non-zero aspect ratios: the difference may be small but non-negligible.

We must also mention at this point the so-called second-order theory proposed by Hudson (1986), which is an attempt at constructing effective properties that depend on linear as well as quadratic terms in the crack density. As has been discussed by Grechka and Kachanov (2006c), this scheme suffers from the shortcoming that it will predict—when starting from some (relatively low) crack density—an increase in stiffness as the crack density increases (at some point exceeding the stiffness of the bulk material), clearly an unphysical behavior.

The choice between various approximate schemes is further hampered by the fact that rigorous *Hashin-Shtrikman bounds* on the elastic constants of materials with inhomogeneities provide no guidance in the case of cracks, for which these bounds degenerate into the trivial statement that cracks will not stiffen the material. This also implies that any of the mentioned approximation schemes is realizable, if the appropriate statistics of mutual positions is chosen, and further emphasizes the importance of the said statistics if interactions are to be taken into account.

Several computational studies have been conducted to examine the effect of interactions, producing somewhat contradictory results. The 2-D simulations of Kachanov (1992), Mauge and Kachanov (1994), Berthaud *et al.* (1994), and Davis and Knopoff (1995) confirm the accuracy of the non-interaction approximation (or, equivalently for cracks, the Mori-Tanaka's scheme) at higher crack densities. Saenger and Shapiro (2002) conclude, however, that the differential scheme is more accurate. 3-D simulations have also produced differing results. The results of Grechka and Kachanov (2006a,c) show that the non-interaction approximation remains accurate at crack densities of at least 0.14. Saenger *et al.* (2004), on the other hand, reached different conclusions that lent support to a differential scheme. Again, when using a NIA model to interpret wave speed data on granite samples deformed at high pressure and high deviatoric stresses, experimental results by Schubnel and Guéguen (2003) suggest that a NIA provides a satisfactory framework to account for the behavior of rocks that contain micro-cracks up to relatively high crack density values of 0.5. For further discussions the reader is referred to papers by Saenger (2007) and by Kachanov (2007).

3.4 Cracks in an Anisotropic Background

Many rocks are anisotropic even if they contain no cracks. This 'background anisotropy' is typically produced by a preferred orientation of crystals or grains (in sedimentary rocks), and/or by a microtexture that develops in the course of some ductile deformation process. An important case is that of shales. Shales represent about 50% of the exposed sedimentary rocks. Apart from their importance to civil engineering works, the properties of shales are of interest where these form cap rocks of hydrocarbon reservoirs or possible sites for storing nuclear waste. The elastic anisotropy of shales can amount to 25% or more in wave speed (Lo *et al.*, 1986; Sarout, 2008a,b) or to 50-60% in elastic stiffnesses. For shales it will often be appropriate to assume a transversely isotropic (TI) matrix. This applies more generally to rocks that possess a foliated structure.

Restricting attention to planar cracks, we now list the consequences for the **B**-tensor (Section 3.1.1), when the response of the background matrix becomes anisotropic:

- The principal directions of **B** generally do not coincide with the crack-normal direction $\mathbf{n}$ and the in-plane directions $\mathbf{s}$ and $\mathbf{t}$ (except for a crack in an orthotropic material normal to one of the orthotropy axes or, in the case of a TI material, parallel or normal to the isotropy plane).

- The crack compliance cannot be decomposed into principal normal and shear components B_N and B_T, respectively, i.e., normal or shear loading of a crack produces *both* a normal and a shear displacement discontinuity (see the results of Mauge and Kachanov, 1994, for the $2-D$ case).
- Crack density parameters, such as the tensors $\boldsymbol{\alpha} = \frac{1}{V}\sum_m (a^3\mathbf{nn})^{(m)}$ and possibly $\boldsymbol{\beta} = \frac{1}{V}\sum_m (a^3\mathbf{nnnn})^{(m)}$, will in general not suffice for characterizing the effective elastic properties. Indeed, such parameters must represent individual cracks in accordance with their actual contribution to an effective compliance. In the case of an anisotropic matrix, these contributions will depend not only on the crack orientation in an arbitrarily chosen reference frame, but also on the orientation of the crack normals $\mathbf{n}$ with respect to the anisotropy axes of the material (a crack that runs perpendicular to a stiff direction of the material produces a higher reduction in the overall compliance than a crack running perpendicular to a soft direction). This difficulty can be avoided in the 2-D case (Tsukrov and Kachanov, 2000), but in 3-D it remains a challenge to be addressed.

For the 2-D case of the orthotropic material, explicit results are available (Mauge and Kachanov, 1994; Tsukrov and Kachanov, 2000). Since they provide insight into the relative importance of various elastic constants of the background and into possible approximations, we shall briefly review these results. For a crack of length $2a$ (the crack line forming an angle φ, $-\pi/2 \leq \varphi \leq \pi/2$, with one of the the principal axes of orthotropy) the components of $\mathbf{B}$-tensor are found to be

$$B_{nn} = C(1+D\cos 2\varphi)a,\ B_{tt} = C(1-D\cos 2\varphi)a,\ B_{tn} = CD(\sin 2\varphi)a. \quad (42)$$

Here the B_{tn} component embodies the shear/normal coupling. The four elastic constants $(E_1^0, E_2^0, G_{12}^0, \nu_{12}^0)$ of the 2-D orthotropic matrix enter through the two constants

$$\begin{aligned} C &= \frac{\pi}{4}\frac{\sqrt{E_1^0}+\sqrt{E_2^0}}{\sqrt{E_1^0E_2^0}}\left(\frac{1}{G_{12}^0}-\frac{2\nu_{12}^0}{E_1^0}+\frac{2}{\sqrt{E_1^0E_2^0}}\right)^{1/2}, \\ D &= \frac{\sqrt{E_1^0}-\sqrt{E_2^0}}{\sqrt{E_1^0}+\sqrt{E_2^0}}. \end{aligned} \quad (43)$$

These results suggest the following simplifications that are also likely to apply–to within a certain degree of approximation–to 3-D problems:

- The strength of the coupling (the ratio of B_{tn} to B_{nn} and B_{tt}) depends on the ratio of Young's moduli E_1^0/E_2^0 and is independent of other matrix moduli. The coupling is weak ($\mid B_{nt} \mid$ is one order of magnitude smaller than $\mid B_{tt} \mid$ and $\mid B_{nn} \mid$) for ratios E_1^0/E_2^0 smaller than 1.4.
- In representing the effective elastic compliances of an anisotropic material with cracks of an arbitrary orientation distribution in the usual manner as a sum
$$S_{ijkl} = S_{ijkl}^0 + \Delta S_{ijkl} \tag{44}$$
of the compliances of the anisotropic matrix and the extra compliance generated by cracks, the latter is determined by placing the cracks in an isotropic background which, in a certain sense, will best approximate the effects of the actual anisotropic matrix (Ougier-Simonin *et al.*, 2009). Such an approximation has been considered in the context of a biomaterial (bone) by Sevostianov and Kachanov (2000, 2001) on the basis of the above 2-D results.

Remark. The question is of course how to find the best isotropic approximation of a given anisotropy. This question was addressed by Fedorov (1968), whose results have been applied in the context of rock mechanics by Arts *et al.* (1991). For the special case of an orthotropic material, the following best-fit expressions (in the sense of an Euclidean norm) for the isotropic bulk and shear moduli, $K_\star$ and $G_\star$ can be given:

$$\begin{aligned} 15G_\star^{-1} &= 4[(1+\nu_{12})E_1^{-1} + (1+\nu_{23})E_2^{-1} + (1+\nu_{31})E_3^{-1}] \\ &\quad + 3(G_{12}^{-1} + G_{23}^{-1} + G_{31}^{-1}), \\ K_\star^{-1} &= (1-2\nu_{12})E_1^{-1} + (1-2\nu_{23})E_2^{-1} + (1-2\nu_{31})E_3^{-1}. \end{aligned} \tag{45}$$

Note that $K_\star$ does not depend on the shear moduli, whereas $G_\star$ depends on all orthotropic constants. This result can be further specialized to a TI material and for the 2-D case.

In 3-D cases, explicit results are available only for a TI material with circular cracks parallel to the isotropy plane (Laws *et al.*, 1985). These results also apply to strongly oblate spheroidal pores with aspect ratios ζ smaller than about 0.1; they have been expressed in terms of crack compliances B_N and B_T by Sarout and Guéguen (2008a,b). The case of spheroids parallel to the symmetry plane of a TI matrix was studied by Nishizawa (1982), using a Differential Effective Method (DEM). This is of interest, because shales can be dealt with in this manner (Sarout and Guéguen, 2008a,b). For 3-D anisotropic materials with *arbitrarily oriented* cracks, no explicit closed form results exist.

3.5 The Effect of Pores

Rocks typically contain a mixture of pores of diverse shapes. For a quantitative characterization of their influence on the overall elastic properties the porosity parameter ϕ will not suffice, even where they preserve overall isotropy: some average shape factor will have to enter into this characterization. The inadequacy of porosity as a sole parameter follows simply from the fact that *two* effective constants can in general not be matched by adjusting *a single* parameter. This also implies that a general isotropic mixture of pores is not equivalent, in its effect on the overall elasticity, to a certain distribution of spherical holes. A simple illustration is provided by the 2-D case of randomly oriented *elliptic holes*, for which an average eccentricity ϵ can be identified explicitly as second parameter, viz.

$$\epsilon = \frac{\pi}{A}\Sigma(a^{(m)} - b^{(m)})^2, \tag{46}$$

where $2a^{(m)}, 2b^{(m)}$ are the axis lengths of the ellipse and A is an averaging area (Kachanov, 1994). This eccentricity vanishes for circles (leaving ϕ as the only parameter), becomes a 2-D crack density for cracks (when ϕ vanishes), and provides a unified coverage of mixtures of diverse eccentricities. The effective bulk and shear moduli are given as functions of ϕ and ϵ by

$$\frac{K}{K_0} = \frac{1}{1 + (2\phi + \epsilon)(1 - \nu_0)^{-1}} \tag{47}$$

and

$$\frac{G}{G_0} = \frac{1}{1 + (4\phi + \epsilon)(2 + 2\nu_0)^{-1}}. \tag{48}$$

This example illustrates the need for a second, shape-sensitive parameter, ϵ, i.e., the effective moduli cannot be expressed solely in terms of ϕ, even in the isotropic case of randomly oriented holes.

For the simplest case of *spherical* pores, the effective elastic constants were obtained by Mackenzie (1950). These results will apply to moderately non-spherical pores, with aspect ratios in the range 0.7 - 1.4, provided the deviations from spheres are randomly oriented and the pores exhibit only moderate surface roughness (in consequence of Hill's comparison theorem, Section 3.2.2). The effective bulk modulus K and shear modulus G are then given in terms of porosity ϕ by the expressions

$$\frac{K_0}{K} = 1 + \frac{3(1 - \nu_0)}{2(1 - 2\nu_0)}\phi \quad \text{and} \quad \frac{G_0}{G} = 1 + \frac{15(1 - \nu_0)}{(7 - 5\nu_0)}\phi, \tag{49}$$

where K_0 and G_0 are the corresponding moduli of the isotropic background matrix, ν_0 being its Poisson's ratio.

For irregularly shaped (non-ellipsoidal) pores, the challenge is to identify those *shape factors* that contribute substantially to the extra compliance generated by pores, and to express this contribution in a sufficiently simple way through appropriate microstructural parameters. Although this problem has not received sufficient attention so far, some progress has been made in this direction, as will now be discussed.

As with irregularly shaped cracks, a general theoretical tool that allows one to deal with irregularly shaped pores is provided by Hill's (1963) comparison theorem. It bounds the contribution of a pore to the overall compliance by the contributions of a circumscribed and an inscribed pore. If ellipsoids are chosen as comparison shapes, then their contributions to the compliance are known and the bounds generated by them can be explicitly calculated. The usefulness of ellipsoids is limited, however, by the widening of the gap between these bounds demanded by pores of complex shape, i.e., concave pores in particular. Hill's bounds can nevertheless provide useful insights. For example, they lead to the conclusion that the following factors will be of minor significance:

- A moderate surface roughness of the pores, i.e. one that can be tightly bounded by smooth shapes. These bounds on the compliance are particularly tight for strongly oblate pores, since for a low aspect ratio, ζ, the pore compliance is quite insensitive to the exact value of ζ.
- The sharpness or bluntness of various corner points, since a sharp corner can be tightly bounded by two blunted corners.

By contrast, the following aspects of pore shape are likely to be important:

- The concavity of a pore. As noted in the Introduction, pores in rocks often have concave shapes. The importance of this factor was demonstrated by explicit calculations for a number of 2-D holes by Zimmerman (1991), who obtained their compressibility (i.e., the change of the pore area under the hydrostatic stress) or the sum ΔS_{iijj} of associated extra compliances, and by Kachanov *et al.* (1994), who examined all components ΔS_{ijkl}. These results show that concave pores are significantly more compliant than the convex pores of the same cross-sectional area and same geometrical symmetry (such as, for example, concave and convex triangles, respectively).
- The interconnectedness of pores (Fig. 6). We distinguish two types of connections: (1) crack-like surfaces and (2) thin pipe-like channels. The two types produce very different effects on the rock elasticity. The first one, from the elasticity viewpoint, is a crack whose compliance scales as its linear size cubed. If this size is comparable to

pore dimensions, as is typically the case, then the effect of the crack is strong and it may overshadow that of the pores connected by this crack. Thin pipe-like channels do not contribute significantly to the overall compliances, as long as their volume fraction remains small. Their effect on overall elasticity is therefore considered to be of minor significance.

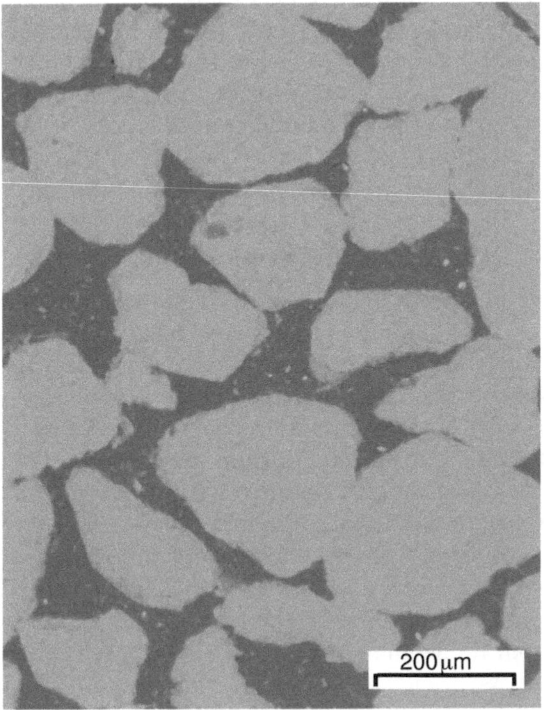

Figure 6. Interconnected pores in Fontainebleau sandstone (SEM) at 25% porosity. (Courtesy J. Fortin).

4 Fluid-saturated Rock with Cracks

The presence of a pore fluid can have a *stiffening effect* on cracks and pores–especially so for low-compressibility liquids–and may hence reduce their contribution to the overall compliance. This calls for an appropriate modification of the crack and pore concentration parameters, since these have to represent individual defects in accordance with their actual contributions to the overall compliance. In the case of cracks, the crack aspect ratio becomes an important factor controlling the stiffening effect of a fluid. In particular, the fourth-rank tensor parameter $\boldsymbol{\beta}$ begins to play an important role; similar parameters will come into play with pores of diverse shapes, in terms of which effective elastic properties can be expressed (Shafiro and Kachanov, 1997). The simplest, albeit somewhat unrealistic case of crack-like pores of identical aspect ratio and random orientation was considered by O'Connell and Budiansky (1974) and Budiansky and O'Connell (1976); in this case, the usual scalar crack density ρ can be retained. In the text to follow, we briefly overview the general case of cracks of diverse orientation distributions and different aspect ratios as well as that of spherical pores. The fluid is assumed to have the constant bulk modulus K_f.

4.1 Effective Elasticity and Poroelasticity

The effective elasticity of fluid-saturated rocks with cracks and pores depends on the rate of loading of an RVE as compared with the rate of pore pressure diffusion (i.e., equilibration) across the RVE. In the context of wave propagation, this implies a frequency dependence of the elastic bulk properties. In order to illustrate this effect, let us consider two limiting cases. In the first an isobaric equilibrium exists throughout the RVE, i.e., the equilibration of the fluid pressure over all interconnected pores and cracks by local flow proceeds at a rate much faster than the rate of loading. In the second limiting case there is insufficient time for significant local flow to take place so that each pore or crack within a given RVE will exhibit its individual pressure response to the (rapid) change in load, as if it were disconnected from its neighbors.

At this point we briefly recall some basic results of Biot's (1941, 1956) theory of poroelasticity (see Chapter 1, and Guéguen *et al.*, 2004, for more detail). A key point is that only the first of the above mentioned limiting cases (isobaric conditions throughout an RVE) lies within the scope of this theory. Indeed, being a macroscopic theory, it treats a porous/cracked rock as a continuous medium, using field variables (as macroscopic point functions) that may be interpreted as appropriate averages over a RVE centered on a point. Pores and cracks are assumed to be interconnected and the pore fluid pressure is assumed to be locally equilibrated over a RVE.

Relaxed (isobaric) compliances, drained and undrained Assuming isobaric conditions for the pore fluid on the scale of an RVE, Biot's theory of poroelasticity admits two types of relaxed compliances, drained and undrained. The *drained* compliances S^d_{ijkl} (in Chapter 1 the superscript 'd' is omitted) are measured in the laboratory on samples that are deformed at a constant fluid pressure by allowing an exchange of pore fluid between the sample and a connected 'reservoir'. But since linear poroelasticity ignores any pressure dependence of elastic constants, drained compliances can be measured on *dry* rock samples, i.e., at atmospheric pore pressures[5]. In field applications, a drained response typically demands a large-scale equilibration of fluid pressures under quasi-static conditions, since the *hydraulic diffusivity* c in most rocks is usually small (a fluid pressure disturbance diffuses through a characteristic distance l in time $\tau \sim l^2/2c$). For example, for a wavelength $l \approx 5$m (frequency in the borehole geophysics kHz range), and a diffusivity $c \approx 10^{-2}\mathrm{m\,s^{-2}}$, as is typical for water in a sandstone of permeability 0.01 darcy and shear modulus of 10 GPa (cf. Guéguen and Palciauskas, 1994), one finds $\tau \approx 10^3$s, which is much larger than a wave period. Thus, no significant pore pressure diffusion will take place across l during a wave period and consequently the drained compliances will not be reflected by wave speeds.

The *undrained* compliances, S^u_{ijkl}, of poroelasticity characterize a deformation in which there is no significant exchange of fluid mass between an RVE and its surroundings, but where the pore fluid pressure is equilibrated on the scale of an RVE (which is thus in an isobaric state) while it may vary from point to point on the macroscopic scale. Such a situation corresponds well to the example considered above, where a perturbation of 5m wavelength propagates through the medium. and were the wavespeeds will 'see' the undrained compliances because no macroscopic fluid flow occurs.

Unrelaxed compliances Compliances measured on a poroelastic solid while the pore pressure varies significantly across an–in this case–*non-isobaric* RVE are referred to as *unrelaxed compliances*; these lie outside the range of applicability of Biot's theory. But the concept of effective elastic properties remains valid, because average stresses and strains can be defined meaningfully in spite of this local non-uniformity in fluid pressure. Fluid pressures induced in pores or cracks now depend on the shape and orientation of these objects, a phenomenon referred to as *pressure polarization.* Above a certain

[5]More accurately, in order to allow for certain chemomechanical interactions between the solid skeleton and the pore fluid, the drained compliances may have to be measured on the 'wetted skeleton' discussed by Biot (cf. Chapter 1).

critical frequency, any experimental measurement of wave speeds will reflect such a non-isobaric state; hence there exists a domain of transition between *low-* and *high-frequency regimes.*

Since the pressure polarization will affect the effective elastic properties, an important observation is that small aspect ratios of pores start to play an important role (in contrast with dry rock, where they play no significant role), because the stiffening effect of the pore fluid depends strongly on the aspect ratio.

Remark. In addition to crack aperture, various irregularities of crack shape that are of little consequence in the dry case, become important as soon as they affect the crack volume, since it is the latter that determines the stiffening effect of the fluid. This implies that, instead of the aspect ratio parameter (the ratio of the maximal aperture of a crack to its diameter), the ratio of the crack volume to its area–i.e. the *average* aperture–is a more pertinent parameter for fluid-saturated, irregularly shaped cracks.

The question is thus at which frequency, f_c, pressure polarization will become noticeable? In fluid saturated, cracked rocks, fluid flow at a micro- or pore-scale is responsible for the frequency dependence of the elastic wave velocities. Ultrasonic measurements in the laboratory provide *high frequency* (MHz) values of the velocities and unrelaxed compliances, whereas field data (as provided by seismological and borehole geophysical surveys) generally fall into the *low frequency* (kHz to Hz) range, characterizing relaxed compliances. As discussed in the above, these low-frequency properties correspond to the relaxed *undrained* quantities of poroelasticity theory. The different flow regimes of the pore fluid that matter in this context are those occurring on the pore-scale within a RVE. The critical frequency f_c is determined by the time needed to equilibrate the fluid pressure between neighbouring cracks by a local *squirt flow* (Le Ravalec *et al.*, 1996):

$$f_c \approx \zeta^3 E_0 / 20\eta. \tag{50}$$

Here the crack aspect ratio $\zeta \approx 10^{-3}$, typically, while Young's modulus $E_0 \approx 20$ GPa for sandstones, and the fluid viscosity $\eta \approx 10^{-3}$ Pa s (water) and $\eta \approx 2 10^{-2}$ Pa s (oil), so that $f_c \approx 1$ kHz or 50 Hz.

4.2 Planar Cracks. Crack-Induced Anisotropy

We now focus on the case of flat and narrow, crack-like pores, and modify the results of Section 3.1 for the case of unrelaxed response. Specializing results obtained by Shafiro and Kachanov (1997) for ellipsoidal fluid-filled pores to the case of a *circular crack* of radius a yields

$$B_T^{sat} = \frac{32(1-\nu_0^2)a}{3\pi E_0(2-\nu_0)}, \qquad \frac{B_N^{sat}}{B_T^{sat}} = (1-\frac{\nu_0}{2})\frac{\delta_f}{1+\delta_f}, \tag{51}$$

as appropriate modification of the result (28) for a dry crack. The parameter δ_f in the second expression characterizes the coupling between the stress and the fluid pressure; it is given by

$$\delta_f = \delta_f^{crack} = \frac{\pi\zeta E_0}{4(1-\nu_0^2)}(\frac{1}{K_f} - \frac{1}{K_0}), \tag{52}$$

and is thus seen to involve the average crack aspect ratio $\zeta = w/a$ (w is the average crack aperture) as an important parameter. Clearly, δ_f will in general be different for different cracks. The parameter δ_f is zero in the limit $\zeta \longrightarrow 0$; this limit corresponds to closed cracks that are allowed to slide without friction. For a highly compressible fluid (air), $\delta_f \longrightarrow \infty$; this limits recovers the case of traction-free cracks (formula (34)). Note that the fluid bulk modulus value K_f is constrained to be smaller than that of the solid matrix K_0. This last restriction holds because of the choice of compliance formulation of the NIA, as explained earlier. This is indeed the situation of fluids in rocks.

In the following, we assume that the values of δ_f and ζ are approximately the same for all cracks. This assumption can actually be weakened. It is enough to assume that the variations in ζ, in crack size, and crack orientation are statistically uncorrelated. It would of course be preferable to employ a distribution of ζ values. However, the lack of appropriate, reliable data has so far prevented a meaningful discussion of this factor. Thus, under the present assumption, equation (34) takes the form

$$\begin{aligned}\Delta S_{ijkl} &= \frac{32(1-\nu_0^2)}{3(2-\nu_0)E_0}\left\{\frac{1}{4}(\delta_{ik}\alpha_{jl} + \delta_{il}\alpha_{jk} + \delta_{jk}\alpha_{il} + \delta_{jl}\alpha_{ik})\right. \\ &\qquad \left. -[1-(1-\frac{\nu_0}{2})\frac{\delta_f}{1+\delta_f}]\beta_{ijkl}\right\}. \end{aligned} \tag{53}$$

In the case of isotropy, i.e. for random crack orientations, $\alpha_{ij} = \frac{1}{3}\rho\,\delta_{ij}$ and one recovers a result of Budiansky and O'Connell (1976). We emphasize that the extra compliances ΔS_{ijkl} (written without a superscript) in (53) are the unrelaxed compliances at high frequencies, as discussed in the above. Here the tensor $\boldsymbol{\beta}$ is seen to enter with the multiplier

$$\psi = 1-(1-\frac{\nu_0}{2})\frac{\delta_f}{1+\delta_f} \tag{54}$$

instead of the factor $\nu_0/2$ for the dry case. Moreover, defining

$$h = \frac{32(1-\nu_0^2)}{3(2-\nu_0)E_0}, \tag{55}$$

the extra compliances (53) become, in more compact form,

$$\Delta S_{ijkl} = h[\frac{1}{4}(\delta_{ik}\alpha_{jl} + \delta_{il}\alpha_{jk} + \delta_{jk}\alpha_{il} + \delta_{jl}\alpha_{ik}) - \psi\beta_{ijkl}]. \tag{56}$$

The fact that the parameter ψ depends on the ratio $K_0\zeta/K_f$ of an 'apparent bulk modulus' of the matrix to the bulk modulus of the pore fluid accounts for the relative importance of the $\boldsymbol{\beta}$-tensor under saturated, unrelaxed conditions. Figure 7 (from Schubnel *et al.*, 2003) provides an illustration for saturated, cracked granite samples; it shows that beyond the onset of dilatancy, as marked by the point C', the components of $\boldsymbol{\alpha}$ and $\psi\boldsymbol{\beta}$ (derived by inversion of 5 velocities)increase substantially. We note here that in the TI case, which is of most interest in geophysics, the number of independent elastic constants is five (for full, non-elliptic TI symmetry), each of them depending on seven parameters (E_0, ν_0, two α components, and three β components).

4.3 Isotropic Mixtures of Spherical Pores

In the simple case of spherical pores the unrelaxed effective bulk and shear moduli are obtained from

$$\frac{K_0}{K} = 1 + \frac{3(1-\nu_0)}{2(1-2\nu_0)}(1 - \frac{1}{1+\delta_f})\phi \quad \text{and} \quad \frac{G_0}{G} = 1 + \frac{15(1-\nu_0)}{(7-5\nu_0)}\phi, \tag{57}$$

where ϕ is the porosity and where the parameter δ_f is given by

$$\delta_f = \delta_f^{sphere} = \frac{2(1-\nu_0)}{(1-2\nu_0)}(\frac{K_0}{K_f} - 1). \tag{58}$$

Note that, as expected, the presence of a fluid does not alter the shear modulus, while it does affect bulk modulus through δ_f. For a highly compressible fluid, such as a gas, δ_f is large and one recovers the result for dry rocks. In the opposite situation, i.e. when the bulk modulus of the fluid is close in value to the bulk modulus of the matrix, the effective bulk modulus K remains almost unchanged, as expected. Relation (57) also shows that the effective bulk modulus exceeds the matrix modulus, if the bulk modulus of the fluid exceeds the latter ($K_f > K_0$). Such a situation is not typical for rocks, however. Note that the compliance formulation is not appropriate in that case.

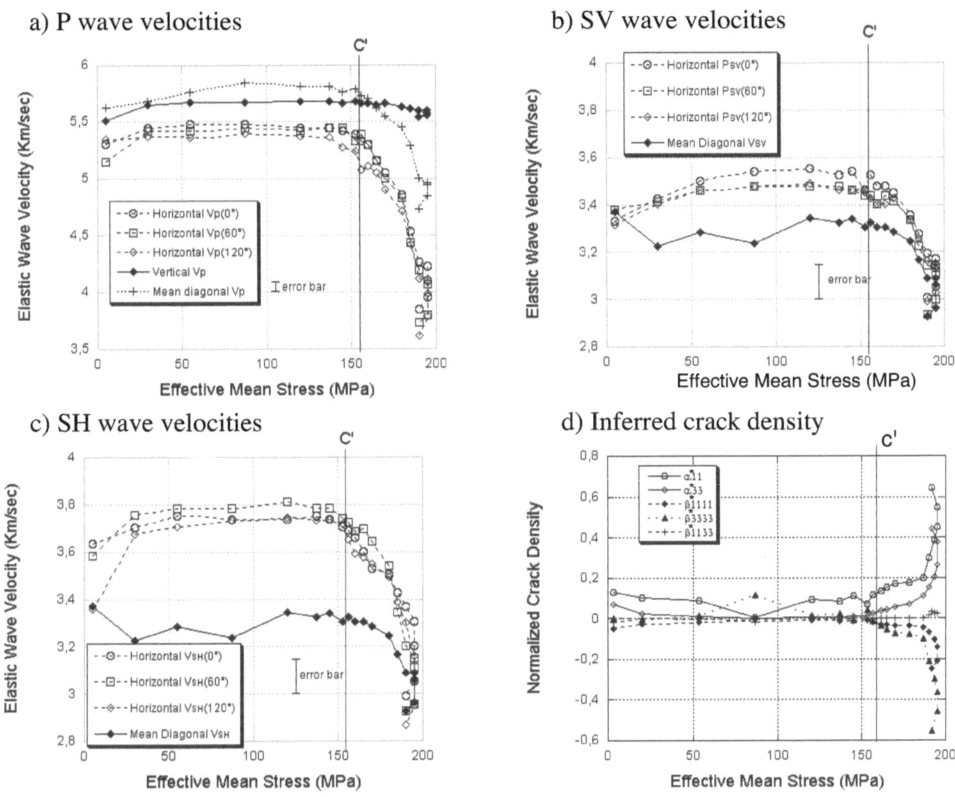

Figure 7. Wet triaxial experiment on Oshima granite at 30 MPa effective confining pressure (confining pressure σ_3 is 40 MPa, and pore pressure P_p is 10 MPa). Effective mean stress is $P' = [(\sigma_1 + 2\sigma_3)/3 - P_p]$. (a) Plain diamonds show vertical V_P, crosses show diagonal V_P; open circles, squares and diamonds show horizontal V_P at $0°, 60°, 120°$ from reference plane respectively; (b) Plain diamonds show diagonal V_{SV}; open circles, squares, and diamonds show horizontal V_{SV} at $0°, 60°$, and $120°$ from reference plane, respectively; (c) Plain diamonds show diagonal V_{SH}, open circles, squares, and diamonds show horizontal V_{SH} at $0°, 60°$, and $120°$ from reference plane, respectively; (d) Normalized crack density tensors versus effective mean stress as inferred from velocity measurements (α_{11} represent the vertical crack density tensor component (Schubnel and Guéguen, 2003).

5 Implications for Wavespeeds

Wave propagation in anisotropic rocks has been a topic of interest for many years, driven by the needs of seismology and seismic prospecting. Wave speeds are measured, both in the laboratory (mostly at MHz frequencies and wavelengths of a few mm) and in the field (at Hz to kHz frequencies and wavelengths from a few meter to a few km). In both cases, wavelengths are larger than pore and crack sizes. Wave speeds provide a way to get important information on rock composition and defect content (pores and cracks, fluid saturation). Laboratory data can provide a reference for the interpretation of field data.

In *dry* rocks, wave speeds are controlled by static effective elastic properties. Both field and laboratory data are in the validity domain of linear elasticity, i.e., both are obtained from low strain amplitude measurements. Both provide static effective properties that can directly be compared.

On the other hand, measurements on *saturated* rocks will reveal a frequency dependence. As discussed in Section 4, the low frequency (quasi-static) regime extends up to a critical frequency f_c in the saturated case. This implies that high-frequency laboratory data cannot be directly compared with low-frequency field data. Poroelastic theory enables us to conduct such a comparison, however, as is discussed in the following.

5.1 Elastic Wave Propagation in a TI Dry Cracked Medium

Thomsen (1986) has introduced three dimensionless parameters $\gamma, \delta, \varepsilon$ that provide a measure of anisotropy for a TI medium. This type of description has subsequently been generalized to orthotropic and monoclinic media by Tsvankin (1997) and by Grechka *et al.* (2000). An examination of the micromechanical information contained in these parameters leads to interesting results, as will be discussed in the following.

Thomsen parameters. The three Thomsen parameters, together with the velocities $V_P^0 = \sqrt{C_{33}/\varrho}$ and $V_S^0 = \sqrt{C_{44}/\varrho}$ characterize seismic anisotropy; the C_{ij} are the elastic constants in the 2-index Voigt notation and ϱ is the bulk density of the rock. For a TI medium this description is complete, since it provides the necessary five parameters. The Thomsen parameters are defined by

$$\varepsilon = \frac{C_{11} - C_{33}}{2C_{33}}, \ \gamma = \frac{C_{66} - C_{44}}{2C_{44}}, \ \delta = \frac{(C_{13} + C_{44})^2 - (C_{33} - C_{44})^2}{2C_{33}(C_{33} - C_{44})}. \quad (59)$$

The parameter combination $\eta = (\varepsilon - \delta)/(1 + 2\delta)$ is called the anellipticity. The Thomsen parameters vanish in the case of isotropy. If all three parameters are much smaller than 1, the anisotropy is considered to be weak. In this case, the angular variations of the phase velocities in any plane containing the symmetry axis x_3 may be approximated by truncating a Taylor expansion in the small quantities $\gamma, \delta, \varepsilon$ after the first order terms, giving

$$\left.\begin{array}{lcl} V_P(\theta) & \approx & V_P^0(1+\delta\sin^2\theta\cos^2\theta+\varepsilon\sin^4\theta) \\ V_{SV}(\theta) & \approx & V_S^0\left(1+\left(\frac{V_P^0}{V_S^0}\right)^2(\varepsilon-\delta)\sin^2\theta\cos^2\theta\right) \\ V_{SH}(\theta) & \approx & V_S^0(1+\gamma\sin^2\theta), \end{array}\right\} \tag{60}$$

where θ is the phase angle between the wave-front normal and the TI axis x_3. There are two shear waves, denoted SH for the pure shear wave polarized perpendicular to x_3 and SV for the pseudo-shear wave polarized normal to the SH wave. Note that for $\eta = 0$ (i.e., $\varepsilon = \delta$) one has $V_{SV}(\theta) \approx V_S^0$ and $V_P(\theta) \approx V_P^0(1+\delta\sin^2\theta)$; in this case any wave front propagating in a plane containing x_3 is elliptic (Thomsen, 1986), $V_P(\theta)$ and $V_{SH}(\theta)$ having the same dependence on $\sin^2\theta$.

Crack-induced transverse isotropy. Consider the type of crack-induced anisotropy that may be characterized approximatively by (36) solely in terms of the symmetric second-rank tensor $\boldsymbol{\alpha}$, the $\boldsymbol{\beta}$-terms being neglected. The overall anisotropy is then an orthotropy of the simplified type, with nine constants reducing to four independent constants that are expressed in terms of two elastic constants of the bulk rock and the three principal components of $\boldsymbol{\alpha}$. In the case of transverse isotropy, $\alpha_{11} = \alpha_{22}$ and the number of independent constants is further reduced to three. This implies that the effective TI constants can be determined from a more limited set of seismic wave data.

We now ask: what relationships exist between the three Thomsen parameters $\gamma, \delta, \varepsilon$ and the four parameters $E_0, \nu_0, \alpha_{11}, \alpha_{33}$? In terms of either E_0, ν_0 or $S_{11}^0 = 1/E_0, S_{12}^0 = -\nu_0/E_0$ (in the Voigt notation), we find that

$$\varepsilon = \delta = \frac{h(\alpha_{33}-\alpha_{11})(S_{11}^0+h\alpha_{11})}{2(S_{11}^0-S_{12}^0+h\alpha_{11})(S_{11}^0+S_{12}^0+h\alpha_{11})}, \tag{61}$$

and

$$\gamma = \frac{h(\alpha_{33}-\alpha_{11})}{4(S_{11}^0-S_{12}^0+h\alpha_{11})}, \tag{62}$$

results, we must note, that are based on the neglect of the $\boldsymbol{\beta}$-terms. This shows that, in the present case, the Thomsen parameters can provide an approximate measure of α_{11} and α_{33}, if the elastic constants of the matrix are known. A further simplification occurs in cases in which either α_{11} or α_{33} vanishes (as, for example, for a single set of parallel cracks).

5.2 Elastic Wave Propagation in a TI Fluid-Saturated Cracked Medium. Frequency Dependence

Depending on the value of the fluid-factor δ_f, the elastic properties of saturated cracked rocks may differ substantially from those of dry rocks in the frequency range (Hz to kHz) of interest to seismology, exploration geophysics, and borehole geophysics (Schubnel and Guéguen, 2003; Dresen and Guéguen, 2004). Two main reasons for this difference are that (1) the $\psi\boldsymbol{\beta}$ term cannot be neglected, and that (2) in fluid-saturated rocks the wave speeds become frequency-dependent.

In the following we shall only consider a TI medium. In this case one has $\alpha_{11} = \alpha_{22}$, $\beta_{1111} = \beta_{2222}$, and $\beta_{1133} = \beta_{2233}$. There are five independent elastic constants that are expressed in terms of two components of $\boldsymbol{\alpha}$, three components of $\psi\boldsymbol{\beta}$, and two matrix constants. Equation (56), together with the general relation for TI symmetry $2\Delta S_{1212} = \Delta S_{1111} - \Delta S_{1122}$ (Nye, 1957) leads to

$$\left.\begin{aligned} \Delta S_{1111} = \Delta S_{2222} &= h(\alpha_{11} - \psi\beta_{1111}) \\ \Delta S_{3333} &= h(\alpha_{33} - \psi\beta_{3333}) \\ \Delta S_{1212} &= h(\tfrac{1}{2}\alpha_{11} - \tfrac{1}{3}\psi\beta_{1111}) \\ \Delta S_{2323} = \Delta S_{3131} &= h(\tfrac{1}{4}(\alpha_{11} + \alpha_{33}) - \psi\beta_{1133}) \\ \Delta S_{1122} &= -\tfrac{1}{3}h\psi\beta_{1111} \\ \Delta S_{2233} = \Delta S_{3311} &= -h\psi\beta_{1133}, \end{aligned}\right\} \tag{63}$$

where ψ and h are defined by (54) and (55), respectively.

Under dynamic loading conditions, the relaxed, low-frequency compliances, S^{lf}_{ijkl}, are equal to the *undrained* compliances (as explained in section 2.1), that is $S^{lf}_{ijkl} = S^{u}_{ijkl}$. Making use of the Brown-Korringa equations of poroelasticity (Brown and Korringa, 1974; see also Section 6 of Chapter 1) these compliances can be related to the drained compliances S^{d}_{ijkl}. The Brown-Korringa equations generalize the well-known Biot-Gassman fluid substitution relation to anisotropic porous media. As shown in Chapter 1, these relationships follow from entirely macroscopic reciprocity arguments and, by making no appeal to any particular pore geometry, must hold universally. As explained in Section 4.1, the drained compliances are usually identified with the dry compliances. The Brown-Korringa equations then furnish the following expression for difference between drained and undrained compliances:

$$S^{d}_{ijkl} - S^{u}_{ijkl} = \frac{[S^{d}_{ijmm} - S^{0}_{ijnn}][S^{d}_{klpp} - S^{0}_{klqq}]}{1/K^{d} - 1/K_0 + \phi(1/K_f - 1/K_0)}. \tag{64}$$

Here the S^{0}_{ijkl} are the matrix rock compliances, K^{d} is the drained bulk modulus, K_f and K_0 are respectively the fluid and solid matrix bulk moduli, ϕ is the rock porosity[6]. In the following, we assume the matrix to be isotropic. In order to obtain the components S^{lf}_{ijkl}, we use equation (64) and take $S^d_{ijkl} = S^0_{ijkl} + \Delta S_{ijkl}$, the extra compliance being given by equation (34). Then

$$\begin{aligned} S^{lf}_{ijkl} &= S^u_{ijkl} \\ &= S^0_{ijkl} + h[\frac{1}{4}(\delta_{ik}\alpha_{jl} + \delta_{il}\alpha_{jk} + \delta_{jk}\alpha_{il} + \delta_{jl}\alpha_{ik}) - \frac{\nu_0}{2}\beta_{ijkl}] \\ &- \frac{h^2(1-\nu_0/2)^2\alpha_{ij}\alpha_{kl}}{1/K^d - 1/K_0 + \phi(1/K_f - 1/K_0)}. \end{aligned} \tag{65}$$

The high-frequency, unrelaxed compliances S^{hf}_{ijkl} now follow from (56):

$$S^{hf}_{ijkl} = S^0_{ijkl} + h[\frac{1}{4}(\delta_{ik}\alpha_{jl} + \delta_{il}\alpha_{jk} + \delta_{jk}\alpha_{il} + \delta_{jl}\alpha_{ik}) - \psi\beta_{ijkl}]; \tag{66}$$

the differences between the high- and low-frequency compliances are thus given by

$$S^{hf}_{ijkl} - S^{lf}_{ijkl} = \frac{h(1-\nu_0/2)}{1+\delta_f}(\alpha_{ij}\alpha_{kl}/\alpha_{mm} - \beta_{ijkl}), \tag{67}$$

taking note of the fact that $1/K^d - 1/K_0 + \phi(1/K_f - 1/K_0) = h(1-\nu_0/2)(1+\delta_f)\alpha_{mm}$, by virtue of (52), (54), (55), and the expression $\phi = \frac{4}{3}\pi\zeta\alpha_{mm}$ for the porosity (Guéguen and Sarout, 2009)

6 Conclusions

The effective elastic properties of rocks, as governed by their internal structure, are of fundamental interest to rock physics, to seismology, to exploration and borehole geophysics, to underground fluid recovery and mining operations, and also many civil engineering works. In seismological applications, this question is of key importance for the interpretation of various elastic waves data, obtained both in the field (at relatively low seismic frequencies, Hz to kHz range) and laboratory data (in a much higher MHz frequency range). Such data may carry useful information on distributions, volume fractions, and fluid content of cracks and pores; a proper interpretation therefore calls for an understanding of the connection between these aspects of the microstructure of a rock and its effective elastic properties.

[6] A somewhat more general interpretation of the quantities S^0_{ijkl}, K_0, and ϕ is made in Chapter 1, Section 6.

Micromechanical analyses allow one to identify those microstructural features that produce a dominant effect on the effective elastic properties and to distinguish them from others that are of minor importance only. Two specific factors that will influence the effect of cracks on the overall elastic properties of a rock are (i) the stiffening effect of a pore fluid and (ii) the often highly irregular geometry of cracks, including crack intersections, wavy cracks, and complex crack orientation distributions. These issues have been dealt with in this review and discussed against a broader background of knowledge available in the mechanics of materials, but limited to linear elastic modeling. This meant, for example, that nonlinearities resulting from crack closure were excluded from consideration. In this regard, the applications contemplated in this review are primarily to wave propagation problems or other situations for which stress changes remain typically too small to cause crack closure.

The main results and conclusions from this survey can be highlighted as follows:

- The cracks, not pores in a rock tend to have the dominant effect on its overall elastic properties.
- The non-interaction approximation (NIA), when formulated in terms of *compliances*, is more broadly applicable than may be expected. A linearization, such that the *stiffnesses* become linear in crack density, leads to a substantial narrowing of the range of applicability of the NIA and constitutes an unnecessary operation.
- Effective elastic properties are generally expressed in terms of second-rank crack density tensor, $\boldsymbol{\alpha} = \frac{1}{V}\sum(a^3\mathbf{nn})^{(m)}$, and fourth-rank tensor $\boldsymbol{\beta} = \frac{1}{V}\sum(a^3\mathbf{nnnn})^{(m)}$ where $a^{(m)}$ is the radius of m-th *circular* crack. These definitions can be extended to flat and intersecting *non-circular* cracks–given their irregularities of shape are randomly distributed–by replacing the distributions of such cracks by equivalent distributions of circular cracks.
- Distributions of *wavy cracks* can be satisfactorily approximated by those of flat cracks, provided the crack shapes do not involve a pattern of multiple waves with high amplitude/wavelength ratios.
- The influence of cracks on the overall elastic properties of *dry rocks* can be characterized, to a first approximation, solely by a symmetric second-rank tensor $\boldsymbol{\alpha}$, as long as the crack surfaces remain close enough to flat.
- The characterization of the effect of dry cracks by the tensor $\boldsymbol{\alpha}$ implies *orthotropy* (orthorhombic symmetry) of the crack-induced anisotropy. Moreover, this orthotropy is of a simple 'elliptic' type, being characterized by four (rather than nine) independent elastic constants that

are expressible in terms of five parameters, i.e., two elastic constants of the rock matrix and three principal values of tensor $\boldsymbol{\alpha}$.

- Cohesive *islands* forming 'welded' or Hertzian contacts on crack faces, even if very small, produce a strong stiffening effect that will reduce the effective crack density.
- The presence of a *pore fluid* may–depending on crack aspect ratios–substantially stiffen a crack. The influence of cracks on the effective elastic properties is then no longer characterized with sufficient accuracy by the second-rank tensor $\boldsymbol{\alpha}$ and the fourth-rank tensor $\boldsymbol{\beta}$ must be taken into account.
- For *fluid-saturated rocks*, the role played by the $\boldsymbol{\beta}$-type tensor implies that the approximation of orthotropy is no longer valid. Because isotropic or transversely isotropic (TI) media are of most interest in geophysical applications, these two cases were dealt with in some detail. For crack-induced TI symmetry it is found that there are 5 independent elastic constants; these incorporate seven parameters, i.e., two matrix elastic constants, two principal values of the tensor $\boldsymbol{\alpha}$ and three components of the tensor $\boldsymbol{\beta}$.
- Given sufficient data, it is possible to extract the crack density parameters from measurements of elastic wave speeds. For *dry rocks* that exhibit crack-induced anisotropy, elastic wave data can provide all four independent elastic constants that describe the effective elastic properties; these depend on two elastic constants of the isotropic (by assumption) rock matrix and on the three principal values of the crack density tensor $\boldsymbol{\alpha}$. If at least one of the two elastic constants of the rock matrix is known, then the components of $\boldsymbol{\alpha}$ can be extracted. For *TI fluid-saturated rocks* there are five independent elastic constants that incorporate seven parameters; additional information is therefore required in this case for extracting the crack density parameters.
- *Frequency* effects arise in saturated rocks as a result of local fluid flow. The frequency dependance of wave speeds (dispersion) can predicted from poroelasticity and effective elasticity models.

Future theoretical work should, in our opinion, focus on further quantitative analyses of structural irregularities in rocks at various scales. Further laboratory studies are needed in order to acquire more data on real rocks in coordination and for the purpose of comparison with theoretical predictions. Experimental low-frequency data on the fluid-saturated rocks are needed in particular. At a larger scale, the construction and use of effective elasticity models for the interpretation of seismic data continues to look promising.

Bibliography

Arts, R.J., Rasolofosaon, P.N.J., and Zinzner, B. (1991). Experimental and theoretical tools for characterizing anisotropy due to mechanical defects in rocks under varying pore and confining pressures. In *Seismic Anisotropy*, edited by Fjaer *et al.*, Society of Exploration Geophysics, Tulsa, Oklahoma, pp. 384–432.

Benveniste, Y. (1986). On the Mori-Tanaka method for cracked solids. *Mechanics Research Communications* 13, 193–201.

Berthaud, Y., Fond, C., and Brun, P. (1994). Effect of interactions on the stiffnesses of cracked media. *Mechanics Research Communications* 21, 525–533.

Biot, M.A. (1941). General theory of three-dimensional consolidation. *Journal of Applied Physics* 12, 155–164.

Biot, M.A. (1956). General solutions of the equations of elasticity and consolidation for a porous material. *Journal of Applied Mechanics* 78, 91–96.

Brace, W.F. (1961). Dependence of fracture strength of rocks on grain size. *Proc. 4th Symp. Rock Mech.*, pp. 99–102, Pennsylvania State University, USA.

Brace, W.F., and E.G. Bombolakis (1963). A note on brittle crack growth in compression. *Journal of Geophysical Research*, 68 (12), 3709-3713, 1963.

Brace, W.F., B.W. Paulding, and C. Scholz (1966). Dilatancy in the fracture of crystalline rocks. *Journal of Geophysical Research* 71, 3939–3953.

Brace, W.F., J.B. Walsh, and W.T. Frangos (1968). Permeability of granite under high pressure. *Journal of Geophysical Research* 73, 2225–2236.

Bristow, J.R. (1960). Microcracks and the static and dynamic constants of annealed and heavily cold-worked metals. *British Journal of Applied Physics* 11, 81–85.

Brown, R.J.S. and J. Korringa (1974). On the dependence of the elastic properties of porous rock on the compressibility of the pore fluid. *Geophysics* 40, 608–616.

Bruner, W.M. (1976). Comment on "Seismic velocities in dry and saturated cracked solids" by R.J. O'Connell and B. Budiansky. *Journal of Geophysical Research* 81, 2573–2576.

Budiansky, B. and R.J. O'Connell (1976). Elastic moduli of dry and saturated cracked solids. *International Journal of Solids Structures* 12, 81–97.

Davis, P.M. and L. Knopoff (1995). The elastic moduli of media containing strongly interacting antiplane cracks. *Journal of Geophysical Research* 100, 18,253–18,258.

Dresen, G. and Guéguen, Y. (2004). Damage and rock physical properties. In *Mechanics of fluid saturated rocks*, edited by Y. Guéguen and M. Boutéca, Academic Press Elsevier, pp. 169-218.

Dyskin, A., Germanovich, L. and Ustinov, K. (1999). A 3-D model of wing crack growth and interaction. *Engineering Fracture Mechanics* 63, 81–110.

Eshelby, J.D. (1957). The determination of the elastic field of an ellipsoidal inclusion and related problems. *Proc. R. Soc. London* A 241, 376-396.

Fedorov, F.I. (1968). *Theory of elastic waves in crystals*. Plenum Press, New York.

Fortin, J., Y. Guéguen, and A. Schubnel (2007). Effect of pore collapse and grain crushing on ultrasonic velocities and Vp/Vs. *Journal of Geophysical Research* 112, B08207, doi:10.1029/2005JB004005.

Grechka, V., P. Contreras, and I. Tsvankin (2000). Inversion of normal moveout for monoclinic media, *Geophysical Prospecting* 48, 577–602.

Grechka, V. and M. Kachanov (2006a). Effective elasticity with closely spaced and intersecting cracks. *Geophysics* 71/3, D85–D91.

Grechka, V., and M. Kachanov (2006b). Seismic characterisation of multiple fracture sets: does orthotropy suffice? *Geophysics* 71/3, D93–D105.

Grechka, V., and M. Kachanov (2006c). Effective elasticity of fractured rocks: A snapshot of work in progress, a Tutorial. *Geophysics* 71, W45-W58.

Grechka, V., I. Vasconselos, and M. Kachanov (2006). The influence of crack shapes on the effective elasticity of fractured rocks. *Geophysics* 71, D153–D160.

Guéguen, Y. and J. Dienes (1989). Transport properties of rocks from statistics and percolation. *Mathematical Geology* 21, 1–13.

Guéguen, Y., T. Chelidze, and M. Le Ravalec (1997). Microstructure, Percolation Thresholds and Rock Physical Properties. *Tectonophysics* 279, 23–35.

Guéguen, Y. and V. Palciauskas (1994). *Introduction to the Physics of Rocks*. Princeton University Press, Princeton, N.J.

Guéguen, Y. and J. Sarout (2009). Crack-induced anisotropy in crustal rocks: predicted dry and fluid-saturated Thomsen's parameters. *Physics of the Earth and Planetary Interiors* 172, 116–124.

Guéguen, Y. and A. Schubnel (2003). Elastic wave velocities and permeability of cracked rocks. *Tectonophysics* 370, 1–4, 163–176.

Guéguen, Y., L. Dormieux, and M. Boutéca (2004). Fundamentals of poromechanics. In *Mechanics of Fluid Saturated Rocks*, edited by Y. Guéguen and M. Boutéca, Academic Press Elsevier, pp. 1–54.

Guéguen, Y. , M. Le Ravalec, and L. Ricard (2006). Upscaling: effective medium theory, numerical methods, and the fractal dream. *Pure and Applied Geophysics* 163, 1175–1192.

Hashin, Z. (1983). Analysis of composite materials - a survey,*Journal of Applied Mechanics*, ASME 50, 481–505.

Hashin, Z. (1988). The differential scheme and its application to cracked materials. *Journal of the Mechanics and Physics of Solids* 36, 719–734.

Hill, R. (1963). Elastic properties of reinforced solids: some theoretical principles. *Journal of the Mechanics and Physics of Solids* 11, 357–372.

Hill, R. (1965). A self-consistent mechanics of composite materials. *Journal of the Mechanics and Physics of Solids* 13, 213–222.

Hood, J.A. and M. Schoenberg (1989). Estimation of vertical fracturing from measured elastic moduli. *Journal of Geophysical Research* 94 (B11), 15611–15618.

Hudson, J.A. (1980). Overall properties of a cracked solid. *Mathematical Proceedings of the Cambridge Philosophical Society* 88, 371–384.

Hudson, J.A. (1986). A higher order approximation to the wave propagation constants for a cracked solid. *Geophysical Journal of the Royal Astronomical Society* 87, 265–274.

Huet, C., P. Navi, and P.E. Roelfstra (1991). A homogenization technique based on Hill's modification theorem. In *Continuum Models and Discrete Systems* 2, edited by by G. Maugin, pp. 135–143.

Jaeger, J.C., N.G.W. Cook, and R.W. Zimmerman (2007). *Fundamentals of Rock Mechanics* (4th edition), Blackwell Publishing Ltd, Oxford, etc.

Ju, J.W. (1990). Isotropic and anisotropic damage variables in continuum damage mechanics. *Journal of the Engineering Mechanics Division*, ASCE 116(12), 2764–2770.

Kachanov, L. (Sr.) (1978). Rupture time under creep conditions.*Izvestia AN SSSR* OTN 8, 26–31. [Republished in *International Journal of Fracture* 97, xi–xviii, 1999.]

Kachanov, M. (1980). Continuum model of medium with cracks. *Journal of the Engineering Mechanics Division* ASCE 106 (EM5), 1039–1051.

Kachanov, M. (1982). A microcrack model of rock inelasticity, parts I and II. *Mechanics of Materials* 1, 19–41.

Kachanov, M. (1987). Elastic solids with many cracks: a simple method of analysis. *International Journal of Solids and Structures* 23, 23–45.

Kachanov, M. (1992). Effective elastic properties of cracked solids: critical review of some basic concepts, *Applied Mechanics Review* 45(8), 304–335.

Kachanov, M. (1994). Elastic solids with many cracks and related problems. In *Advances in Applied Mechanics*, edited by J. Hutchinson and T. Wu, Academic Press, pp. 256-426.

Kachanov, M. (1998). Solids with cracks and pores of various shapes: proper parameters of defect density and effective elastic properties. *International Journal of Fracture* 88, 1–31.

Kachanov, M. (2007). On the effective elastic properties of cracked solids - Editor's comments. *International Journal of Fracture*146, 295–299.

Kachanov, M., I. Tsukrov, and B. Shafiro (1994). Effective moduli of solids with cavities of various shapes. *Applied Mechanics Reviews* 47(1), S151–S174.

Kachanov, M. and I. Sevostianov (2005). On quantitative characterization of microstructures and effective properties. *International Journal of Solids and Structures* 42, 309–336.

Kachanov, M., R. Prioul, and J. Jocker (2010). Incremental linear elastic response of rocks containing multiple rough fractures: similarities and differences with traction-free cracks. *Geophysics* 75(1), D1–D11.

Kanaun, S.K. and V.M. Levin (2008). *Self-consistent Methods for Composites*, Springer-Verlag, 2008.

Kunin, I.A. (ed.) (1983). *Elastic media with microstructure* Vol.2, Springer-Verlag.

Landau, L. and E. Lifchitz (1967). Théorie de l'Élasticité, Mir Editors, Moscou.

Law, R.D. (1990). Crystallographic fabrics: a selective review of their applications to research in structural geology. In *Deformation Mechanisms, Rheology and Tectonics*, edited by R.L. Knipe *et al.*, Geological Society Special Publication No. 54, The Geological Society London, pp. 335–352.

Laws, N., G. Dvorak, and M. Hejazi (1983). Stiffnesses changes in unidirectional composites caused by crack systems, *Mechanics of Materials*, 2, 123-137, 1983.

Lehner, F.K. and M. Kachanov (1996). On modelling of 'winged' cracks forming under compression. *Int. Journal of Fracture* 77, R69–R75.

Lekhnitsky, S.G. (1963). *Theory of Elasticity of an Anisotropic Elastic Body*. Holden-Day, San-Francisco.

Le Ravalec, M. and Y. Guéguen (1994). Permeability models for heated saturated igneous rocks. *Journal of Geophysical Research* 99, 24,251–24,261.

Le Ravalec, M. and Y. Guéguen (1996a). High and low frequency elastic moduli for saturated porous/cracked rock (differential, self consistent and poroelastic theories). *Geophysics* 61, 1080–94.

Le Ravalec, M. and Y. Guéguen (1996b). Comments on "The elastic moduli of media containing strongly interacting antiplane cracks" by P.M. Davis and L. Knopoff. *Journal of Geophysical Research* 101, 25,373–25,375.

Le Ravalec, M., Y. Guéguen, and T. Chelidze (1996). Elastic waves velocities in partially saturated rocks. *Journal of Geophysical Research* 101, 837–844.

Mackenzie, J.K. (1950). The elastic constants of solids containing spherical holes. *Proc. R. Soc. London* 63B, 2–11.

Madden, T.R. (1983). Microcrack connectivity in rocks: A renormalization group approach to the critical phenomena of conduction and failure in crystalline rocks. *Journal of Geophysical Research* 88 (B1), 585-592.

Markov (2000). Elementary micromechanics of heterogeneous media. In: *Heterogeneous Media - Micromechanics Modeling Methods and Simulation*, edited by K. Markov and L. Preziosi, Birkhäuser, Basel, pp. 1–162.

Mauge, C. and M. Kachanov (1994). Effective elastic properties of an anisotropic material with arbitrarily oriented interacting cracks. *Journal of the Mechanics and Physics of Solids* 42, 561–584.

Mavko, G. and A. Nur (1978). The effect of nonelliptical cracks on the compressibility of rocks. *Journal of Geophysical Research* 83 (B9), 4459–4468.

Mear, M., I. Sevostianov, and M. Kachanov (2007). Elastic compliances of non-flat cracks. *International Journal of Solids and Structures* 44, 6412–6427.

Mori, T. and K. Tanaka (1973). Average stress in matrix and average elastic energy of materials with misfitting inclusions. *Acta Metallurgica* 21, 571–574.

Mukerji, T., Berryman, J., G. Mavko, and P. Berge (1995). Differential effective medium modelling of rock elastic moduli with critical porosity constraints. *Geophysical Research Letters* 22(5), 555–558.

Nasseri, M.H.B., Schubnel, A., and Young, R.P. (2006). Coupled evolutions of fracture toughness and elastic waves properties at high crack density in thermally treated Westerly granite, *Int. J. Rock Mechanics and Mining Sciences*, doi:10.1016/j.ijrmms.2006.09.008.

Nishizawa, O., Seismic velocity anisotropy in a medium containing oriented cracks -Transversely isotropic case, *Journal of Physics of the Earth*, 30, 331-347, 1982.

Nye, J.F. (1957). *Physical Properties of Crystals*. Oxford University Press, Oxford etc., (reprinted in 1979).

O'Connell, R.J. and B. Budiansky (1974). Seismic velocities in dry and saturated cracked solids. *Journal of Geophysical Research* 79, 5412–5426.

Ougier-Simonin, A., J. Sarout, and Y. Guéguen (2009). A simplified model of effective elasticity for anisotropic shales. *Geophysics*, 74, 3, D57-D63.

Piau, M. (1980). Crack-induced anisotropy and scattering in stressed rocks. *International Journal of Engineering Science* 18, 549–568.

Rice, J.R. (1979). Theory of precursory processes in the inception of earthquake rupture. *Gerlands Beitrage Geophysik* 88, 91-121.

Saenger, E.H. and S.A. Shapiro (2002). Effective velocities in fractured media: a numerical study using the rotated staggered finite-difference grid. *Geophysical Prospecting* 50, 183–194.

Saenger, E.H. (2007). Comment on "Comparison of the non-interaction and the different schemes in predicting the effective elastic properties of fractured media" by V. Grechka. *International Journal of Fracture* 146, 291–292.

Saenger, E.H., O.S. Krüger, and S.A. Shapiro (2002). Effective elastic properties of randomly fractured soils: 3-D numerical experiments. *Geophysical Prospecting* 52, 183–195.

Sarout, J. and Y. Guéguen (2008a). Anisotropy of elastic wave velocities in deformed shales. Part I: Experimental results. *Geophysics* 73/5, D75–D89.

Sarout, J. and Y. Guéguen (2008b). Anisotropy of elastic wave velocities in deformed shales. Part II: Modeling results. *Geophysics* 73/5, D91–D103.

Sayers, C.M. and M. Kachanov (1995). Microcrack Induced Elastic Wave Anisotropy of Brittle Rocks. *Journal of Geophysical Research* 100, 4149–4156.

Schoenberg, M. (1980). Elastic wave behavior across linear slip interfaces. *Journal of the Acoustical Society of America* 68, 1516–1521.

Schoenberg, M. and J. Douma (1988). Elastic wave propagation in media with parallel fractures and aligned cracks. *Geophysical Prospecting*, 36, 571–590.

Schubnel, A. and Y. Guéguen (2003). Dispersion and anisotropy of elastic waves in cracked rocks. *Journal of Geophysical Research* 108, 2101–2116.

Schubnel, A., O. Nishizawa, K. Masuda, X.J. Lei, Z. Xue, andY. Guéguen (2003). Velocity Measurements and Crack Density Determination during Wet Triaxial Experiments on Oshima and Toki Granites. *Pure and Applied Geophysics* 160, 869–888.

Sevostianov, I. and M. Kachanov (2000). Impact of the porous microstructure on the overall elastic properties of the osteonal cortical bone. *Journal of Biomechanics* 33, 881–888.

Sevostianov, I. and M. Kachanov (2001a). Author's response. *Journal of Biomechanics* 34, 709–710.

Sevostianov, I. and M. Kachanov (2001b). Elastic compliance of the annular crack. *International Journal of Fracture* L51–L54.

Sevostianov, I. and M. Kachanov (2002). On the elastic compliances of irregularly shaped cracks. *International Journal of Fracture* 114, 245–257.

Sevostianov, I. and M. Kachanov (2008a). Contact of rough surfaces: A simple model for elasticity, conductivity, and cross-property connections. *Journal of Mechanics and Physics of Solids* 56, 1380–1400.

Sevostianov, I. and M. Kachanov (2008b). Normal and tangential compliances of interface of rough surfaces with contacts of elliptic shape. *International Journal of Solids and Structures* 45, 2723-2736.

Shafiro, B. and M. Kachanov (1999). Solids with non-spherical cavities: simplified representations of cavity compliance tensors and the overall anisotropy. *Journal of The Mechanics and Physics of Solids* 47, 877–898.

Shafiro, B. and M. Kachanov (1997). Materials with fluid-filled pores of various shapes: effective moduli and fluid pressure polarization, *International Journal of Solids and Structures* 34, 3517–3540.

Simmons, G. and W.F. Brace (1965). Comparison of static and dynamic measurements of compressibility of rocks. *Journal of Geophysical Research* 70, 5649–5656.

Simmons, G. and D. Richter (1976). Microcracks in rocks. In *The Physics and Chemistry of Minerals and Rocks*, edited by R.G.J. Strens, John Wiley, New York, pp. 105–137.

Thomsen, L. (1986). Weak elastic anisotropy. *Geophysics*, 51/10, 1954-1966.

Tsukrov, I. and M. Kachanov (2000). Effective moduli of an anisotropic material with elliptical holes of arbitrary orientational distribution.*International Journal of Solids and Structures* 37, 5919–5941.

Tsvankin, I. (1997). Anisotropic parameters and P-wave velocity for orthorhombic media. *Geophysics* 62, 1292–1309.

Vavakin, A.S. and R.L. Salganik (1975). Effective characteristics of nonhomogeneous media with isolated inhomogeneities. *Mechanics of Solids*, Allerton Press, 58–66. (English translation of *Izvestia AN SSSR, Mekhanika Tverdogo Tela* 10, 65–75).

Walpole, L. (1984). Elastic behavior of composite materials: theoretical fundations. *Advances in Applied Mechanics* 21, 168–242.

Walsh, J. B. (1965a). The effect of cracks on the compressibility of rocks. *Journal of Geophysical Research* 70, 381–389.

Walsh, J. B. (1965b). The effect of cracks on uniaxial compression of rocks. *Journal of Geophysical Research* 70, 399–411.

Zoback, M.D. and J.D. Byerlee (1975). The effect of microcrack dilatancy on the permeability of Westerly granite. *Journal of Geophysical Research* 80, 752–755.

Zimmerman, R. (1991). *Compressibility of Sandstones*. Elsevier, Amsterdam.

Three-dimensional morphology evolution of solid-fluid interfaces by pressure solution

Jean L. Raphanel

Laboratoire de Mécanique des Solides, CNRS,
Ecole Polytechnique, Palaiseau, France

Abstract This chapter offers an introduction to recent theoretical research on chemo-mechanical phenomena in solid/liquid systems under non-hydrostatic stress. Its purpose is to acquaint the reader with a number of key concepts and results through an analysis of one specific problem. The problem chosen is the surface roughness of a stressed elastic solid that is dissolving in a solution phase, as determined by the state of stress in the solid. This roughness is treated as a surface instability that is governed by a three-dimensional, local criterion derived from a linear stability analysis of a homogeneously stressed half-space. Using a normal mode decomposition, with modes in the shape of a polarized wave, the solution to the linear elasticity problem is constructed by superposition of three Galerkin vectors to account for the three components of the stress vector acting on the surface. It is shown that for isotropic surface properties, the dominant mode determining the instability has its normal parallel to the maximum principal stress. The surface morphology evolution is thus shown to be controlled by the principal stress directions. The three-dimensional, local stability criterion is then applied to experimental results reported by den Brok and Morel for single-crystal K-alum specimens in the shape of a rectangular plate with a cylindrical hole that allowed the creation of a stress gradient. The wavy pattern, which is observed in these experiments only under an applied load, is well predicted when interpreting it theoretically as a surface instability. However, the wavelength of the dominant mode of instability was expected theoretically to vary by several orders of magnitude due to the stress gradient across the specimen. This dependence, and the predicted growth in time of the instability has not been observed in the laboratory, suggesting further complexity in the stress-driven evolution of surface morphologies, which remains to be explored both theoretically and experimentally.

1 Introduction

In this chapter we wish to offer the reader a glimpse into the chemomechanics of solid/liquid systems under non-hydrostatic stress, a subject that is receiving increasing attention in the study of diagenetic processes and rock deformation mechanisms. Its theoretical foundations were laid by J.W. Gibbs (1878) in his classic memoir "On the Equilibrium of Heterogeneous Substances". From these principles it has long been apparent that non-hydrostatically stressed solid/liquid interfaces in rocks are rarely, if ever, in a stable thermodynamic equilibrium state. The question of interfacial stability and that of the morphological evolution of an interface and its underlying kinetics thus arise. The scope of this area of study is further enlarged by the occurrence of processes such as *intergranular pressure solution*, which in essence can be seen as a complex chemomechanical contact problem for wet grain boundaries.

The theoretical tools needed for dealing with these questions have been developed over the last four or five decades. It is the purpose of this chapter to acquaint the reader with these key concepts by way of an example : the analysis of the surface roughness of a stressed elastic solid that is brought in contact with an aqueous solution phase. Here the surface morphology is considered as resulting from a surface instability that is governed by a three-dimensional, local criterion derived from a linear stability analysis. The predictions of our theoretical analysis are then compared with experimental observations made by den Brok and Morel (2001) on initially plane specimens that were found to develop surface roughness when loaded in the elastic range. Their specimens consisted of K-alum salt single crystals in the shape of a rectangular plate with a cylindrical hole which allowed the creation of a stress gradient (Morel and den Brok, 2001).

Den Brok and Morel attempted to explain the wavy pattern developing under an applied load on the surfaces of their specimens by appealing to an analysis of Srolovitz (1989), who studied two-dimensional surface instabilities driven by a surface diffusion process. The wavelengths predicted in this manner compared favorably with the observed pattern although their expected dependency on stress (proportionality to $1/\sigma^2$) could not be clearly established.

Here we shall follow a different avenue to obtain an exact, linear criterion for a surface instability that is driven by pressure solution (rather than surface diffusion) on the surface of an elastic solid in a homogeneous, non-hydrostatic state of stress. The predictions made by this new, local criterion will be then be compared with the experimental results of den Brok and Morel (2001).

This chapter has been written with the didactic purpose of bringing together three important concepts or elements in one particular study. First, pressure solution as a deformation mechanism is discussed from a theoretical point of view, putting emphasis on thermodynamic arguments that permit an identification of the driving forces at work in this compound solution-diffusion-precipitation process. The relevance of experiments such as those of den Brok and Morel (2001) in clarifying the fundamentals of dissolution processes along stressed grain boundaries is emphasized in this context. The second concept, which is exposed in section 3, is that of a linear stability analysis for a flat solid/fluid interface, when the non-hydrostatic stress in the solid is in the elastic range. This exercise provides an opportunity for discussing a linearization technique with many applications in solid mechanics. As a novelty, we shall consider surfaces in three-dimensions, which will allow us to study the orientation of a wavy instability with respect to the principal stress directions in the solid. The third element is an analytic solution of the elasticity problem, involving the superposition of three Galerkin vectors, each corresponding to a component of the stress vector acting on the interface. In section 4 we prove that the dominant mode of instability is indeed oriented perpendicular to the maximum compressive stress, thereby justifying the application of a 2-D analysis for isotropic, linear elastic solid phases and isotropic surface effects.

We then propose to apply the results of the stability analysis of a homogeneously stressed half space as a local criterion to 3-D problems that involve a stress gradient. This is the case of the specimen tested by den Brok and Morel (2001) and studied in section 5. The stress distribution produced by the loading is computed with a finite-element approximation, assuming plane stress. The local criterion is applied to the stress state determined at each reduced-integration point of the four-noded elements considered for spatial discretization. It is shown that the observed spatial variation of the orientation of the wavy pattern that develops under stress is indeed controlled by the principal compressive stress direction. However, the sensitivity of the wavelength to the stress magnitude and the wave amplitudes, because of varying instability rate of growth, do not agree with laboratory observations. These differences suggest that the physics involved in these experiments are not fully described by the proposed theory.

2 Pressure Solution as a Deformation Mechanism

Pressure solution is a deformation mechanism that is characterized by the dissolution of minerals within stressed grain contacts, the local transport of the dissolved species along these contacts toward low-stress surfaces (e.g., pore walls) and/or the advective removal of dissolved material by a percolating aqueous pore fluid. The last element of large-scale transport necessitates an open-system description of pressure solution (Lehner, 1995; Lehner and Leroy, 2004). If absent, a closed-system description will be appropriate and pressure solution as a deformation mechanism then resembles Coble creep (Coble, 1963) from which it differs mainly by the presence of a fluid phase within grain boundaries (Rutter, 1976, 1983). Pressure solution results in the lithification and compaction of sedimentary rocks and low grade metamorphic rocks and is thus a key factor controlling the quality of potential hydrocarbon reservoirs. Pressure solution is also believed to promote the healing of freshly ruptured shear zones, explaining their strength recovery (Sleep, 1995). The same process is observed to operate at high rates in halite and other highly soluble salts and these are therefore often used as analogue materials in experimental studies (Urai et al., 1986; Spiers et al., 1990; Morel and den Brok, 2001).

Pressure solution is activated or enhanced by stress concentrations as exist at grain boundaries (Fig. 1a). Observing the ongoing process in the laboratory is difficult and requires a dedicated set-up (see, e.g., Schutjens and Spiers, 1999). It has been suggested that the grain boundary is composed of fluid-infiltrated, interconnected cavities often described as a network of islands and channels, as illustrated schematically by the cross-section shown in Figure 1b (Raj, 1982; Lehner and Bataille, 1984/85; Spiers and Schutjens, 1990). Such an interconnected network of channels provides a path for the diffusive transport of dissolved material to the pore space. An alternative grain boundary model (Weyl, 1959; Rutter, 1976) assumes the presence of a thin, uniform film of a 'grain boundary fluid', possibly in an altered state, but again providing a path for the relatively rapid diffusive transport of solute components. While it is difficult to maintain such a picture for realistic contacts between the individual grains of a sandstone, for example, which often appear very irregular and sutured (see, e.g., Houseknecht, 1987; Houseknecht and Hathon, 1987), thin film models might well offer an adequate description in certain circumstances, typically of an idealized kind. They are therefore still debated in the literature (see, e.g., Hickman and Evans, 1995).

Thin film models do however raise the question of the stability of a film respectively that of its break-up, followed by an evolution at the micron scale of the grain boundary morphology towards an island/channel structure. The initial geometry of the film, at a small scale, can be idealized as a half-space in contact with the aqueous solution phase, as illustrated in Fig. 1c. It is to such questions that the experiments performed by den Brok and Morel (2001) on analogue materials at the more accessible millimeter scale can

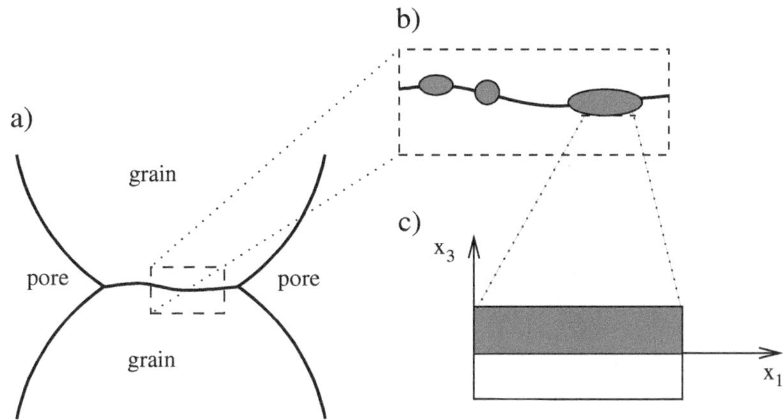

Figure 1. a) Two grains in contact deforming by pressure solution; b) grain boundary with islands-channel structure; c) grain boundary fluid forming flat, film-like channel raises question of morphology evolution by dissolution and precipitation, motivating the 3-D boundary-value problem illustrated by Figure 2.

contribute.

The key theoretical elements called for in modeling pressure solution are a condition governing the thermodynamic equilibrium of a stressed solid in contact with its solution and, secondly, an appropriate theoretical description of the stress-sensitive kinetics of interface propagation. These are introduced first in the following two sections.

2.1 Conditions of Equilibrium of a Pure Solid in Contact with a Fluid Solution

The conditions governing the thermodynamic equilibrium between a stressed solid and its fluid solution have first been studied and formulated by Gibbs (1878). The system solid-solution is in a state of thermodynamic equilibrium, if it satisfies appropriate conditions of mechanical, chemical, and thermal equilibrium. In the following, the last condition will be taken as fulfilled in all cases, assuming that isothermal conditions prevail at all times throughout the entire system. It suffices therefore to deal with the conditions of mechanical and chemical equilibrium.

For the solid phase to be in mechanical equilibrium throughout a region

Ω_s, we require

$$\operatorname{div} \boldsymbol{\sigma}_s = \mathbf{0} \quad \forall \mathbf{x} \in \Omega_s \,, \tag{1}$$

assuming no body forces to come into play. Here the subscript s refers to the solid phase.

The fluid phase is always assumed to be in a state of hydrostatic equilibrium at a uniform isotropic stress $\boldsymbol{\sigma}_f = -p_f \boldsymbol{\delta}$, where $\boldsymbol{\delta}$ is the identity tensor.

The express the condition of local mechanical equilibrium at any given point along the solid-solution interface, we consider the vector $\mathbf{T}^s_{(\mathbf{n})} = \boldsymbol{\sigma}_s \cdot \mathbf{n}$ defining the action exerted by the solid phase on the 'solid side' of the interface (with unit normal $\mathbf{n}$ pointing into the solid phase), and the vector $\mathbf{T}^f_{(-\mathbf{n})} = -p_f \boldsymbol{\delta} \cdot (-\mathbf{n})$ defining the action by the fluid phase on the 'fluid side' of the interface (with unit normal $-\mathbf{n}$). The solution surface is treated here a surface of discontinuity that carries its own 'excess densities' of energy and entropy, the existence of which demands a certain amount of work be spent in the formation of new surface, work that must in general be distinguished from work spent in stretching by a work-conjugate membrane stress (Gibbs, 1878, p. 315), although for isotropic solids the resultant conditions of interfacial mechanical equilibrium will be indistinguishable and are given by (see also Alexander and Johnson, 1985; Heidug, 1991)

$$\mathbf{T}^s_{(\mathbf{n})} = -\mathbf{T}^f_{(-\mathbf{n})} + 2H\gamma\, \mathbf{n} \,, \tag{2}$$

implying the existence of an interfacial force density proportional to an apparent 'interfacial tension' γ (in Pa m) and the mean curvature $2H = 1/r_1 + 1/r_2$ - *positive* when the centers of the principal radii of curvature r_1 and r_2 lie on the same side as the fluid (Gibbs made use of the opposite sign convention). The external tractions $\mathbf{T}^s_{(\mathbf{n})}$ and $\mathbf{T}^f_{(-\mathbf{n})}$ on the two sides of the interface are thus balanced by an appropriate internal force density.

¿From (2) the stress vectors in the two phases on either side of the interface are seen to suffer a 'jump' in their normal component only, their tangential components vanishing on either side, i.e.,

$$\mathbf{n} \cdot \boldsymbol{\sigma}_s \cdot \mathbf{n} = 2H\gamma - p_f, \quad \text{and} \quad \mathbf{t} \cdot \boldsymbol{\sigma}_s \cdot \mathbf{n} = -p_f \mathbf{t} \cdot \mathbf{n} = 0, \tag{3}$$

where $\mathbf{t}$ is any vector tangent to the interface.

We note that when the solid phase sustains but an isotropic stress $-p_s\boldsymbol{\delta}$, condition (3) reduces to Laplace's condition $p_f - p_s = 2H\gamma$.

The condition of local interfacial thermodynamic equilibrium between a pure solid under nonhydrostatic stress and its (e.g. aqueous) solution was first obtained by Gibbs (1878, eq. 661). Gibbs' result may be put in the form

$$\psi_s + \frac{p_f - 2H\gamma}{\rho_s} = \mu_S\,, \tag{4}$$

where μ_S is the mass-specific chemical potential of the solute (subscript upper case S not to be confused with the lower case s referring to the solid), ψ_s denotes the specific Helmholtz free energy, and ρ_s the mass density of the solid phase.

It is interesting to note that whenever (2) applies, the left-hand side of (4) may be interpreted as the normal component of a "chemical potential tensor" of the pure solid constituent $\boldsymbol{\mu}$ (Grinfeld, 1982, 1991; Truskinovsky, 1984; Heidug and Lehner, 1985; Lehner, 1995, 1997), as defined by[1]

$$\boldsymbol{\mu} = \psi_s \boldsymbol{\delta} - \boldsymbol{\sigma}_s/\rho_s.$$

If we write

$$\boldsymbol{\mu} = \mu_S \boldsymbol{\delta},$$

for the isotropic chemical potential of the solute in solution, which appears on the right-hand side of (4), then this condition is seen to express a continuity requirement on the normal component of the chemical potential of a species that is present as a pure solid on one side of the phase boundary and as dissolved solute on the other. Equation (4) may thus be written

$$\mathbf{n} \cdot [\![\boldsymbol{\mu}]\!] \cdot \mathbf{n} = 0,$$

making use of the "jump bracket" notation $[\![\boldsymbol{\mu}]\!] = \boldsymbol{\mu}^+ - \boldsymbol{\mu}^-$ and interpreting the chemical potential in accord with the phases present on the positive and negative side of the phase boundary (the surface normal $\mathbf{n}$ always points into the phase on the positive side of the interface).

Under isothermal conditions, the Helmholtz free energy density ψ_s in (4) of an isotropic elastic solid is given by the strain energy density[2]

$$\psi_s = \frac{1}{2\rho_{s0}} \boldsymbol{\sigma} : \boldsymbol{\epsilon} = \frac{1}{2\rho_{s0} E} \left[(1+\nu) \mathrm{tr}\, \boldsymbol{\sigma}_s^2 - \nu \left(\mathrm{tr}\, \boldsymbol{\sigma}_s \right)^2 \right]. \tag{5}$$

[1] Somewhat earlier, Bowen and Wiese, 1969, and Bowen, 1976, had introduced a chemical potential tensor in a continuum theory of fluid mixtures. However, non-hydrostatic stresses are dissipative and arise with non-equilibrium processes in their theory; consequently the role played by a chemical potential tensor in solid/solid or solid/liquid phase equilibria lies beyond its scope.

[2] The double-dot denotes the scalar product of two second order tensors, giving $\mathbf{A}:\mathbf{B} = A_{ij}B_{ij}$ in a Cartesian coordinate system; 'tr' is the trace operator, giving $\mathrm{tr}\,\mathbf{A} = A_{ii}$; summation over repeated indices being carried out in all cases.

Here ρ_{s0} is the density of the solid in a reference state, ν is Poisson's ratio, and E Youngs modulus. The reference density in (5) and current density in (4) are related by

$$\frac{\rho_{s0}}{\rho_s} = 1 + \frac{1-2\nu}{E} \mathrm{tr}\, \boldsymbol{\sigma}_s \,. \tag{6}$$

The distinction between the current and reference material densities will be kept in the linearized stability analysis presented in the next section because close to thermodynamic equilibrium the relative variation in material density $(\rho_s - \rho_{s0})/\rho_{s0}$ is of the same order as the relative variation of the Helmholtz free energy, as discussed in Leroy and Heidug (1994). This results holds even under the assumption of small perturbations, i.e., infinitesimal deformation, small rotation and small displacement compared to the characteristic length of the problem.

2.2 Kinetics of the Interface Propagation

This kinetics sets in the only time parameter of this problem and is thus central to the stability argument. The migration of the interface corresponds to a flux of material J_N which is positive for dissolution and negative for deposition. The material flux is related to the Lagrangian speed of propagation S_N

$$J_N = -\rho_{s0} S_N \,, \tag{7}$$

disregarding any change in the surface area of the interface under stress. The velocity is negative for a positive flux, since the interface is migrating towards the solid phase during dissolution. The thermodynamic force conjugate to the mass flux is denoted χ and is derived by application of the thermodynamics of irreversible processes (De Groot and Mazur, 1962) already considered in physical metallurgy by Machlin (1953). For pressure solution, it has been shown (Lehner and Bataille, 1984; Heidug and Leroy, 1994), from an estimate of the dissipation during migration, that the driving force is the left-hand side of equation (4)

$$\chi \equiv \psi_s + \frac{p_f - 2H\gamma}{\rho_s} - \mu_S \,. \tag{8}$$

The kinetic law controlling the interface migration is thus expressed by any function $\mathcal{L}(\chi)$ having the following properties

$$J_N = \mathcal{L}(\chi) \quad \text{with} \quad \mathcal{L}(\chi)\chi \geq 0 \quad \forall \chi \,. \tag{9}$$

The simplest expression for $\mathcal{L}$ is the linear function

$$\mathcal{L}(\chi) = L\chi \,, \tag{10}$$

in which the phenomenological constant L may be inferred from experimental data (see, for instance, Rimstidt and Barnes, 1980). We shall retain this form in the following but we also note that the choice of the same constant for dissolution and precipitation, although of interest for the simplicity of the theoretical development, is certainly incorrect for some materials like, for instance, quartz (cf. the discussion in Lehner and Leroy, 2004). It may also bear some consequences on the validity of our interpretation of den Brok and Morel (2001) experiments, albeit on yet another type of material, as will be discussed in the last section.

3 Linear Stability

The objective of this section is to present in a didactic way the different steps leading to solving the linear stability problem for the thermodynamic equilibrium of the elastically stressed solid dissolving in an aqueous solution phase. The first step is the description of the infinitesimal geometric perturbation of the otherwise flat solid–fluid interface. The general structure of the perturbation in then extended to all relevant field variables. The linearization of the Laplace boundary conditions sets the first order problem. This three dimensional elasticity problem has then to be solved by an appropriate scheme. Here, we construct a solution, that is a stress field satisfying mechanical equilibrium and boundary conditions, by the superposition of Galerkin vectors (Fung,1965).

3.1 Perturbation of the Interface

The solid-fluid interface is assumed to form a flat surface in the (1,2) plane (Figure 2). Let this surface be subjected now to a perturbation of infinitesimal amplitude in the third coordinate direction. The local surface normal points into the fluid phase. A point $\mathbf{y}$ on this surface has the coordinates (y^1, y^2, y^3) in the orthonormal basis $\{\mathbf{e}_i\}$, the third contravariant coordinate being given by

$$y^3 = 0 + \varepsilon Y f(\mathbf{y}, t)\,, \quad \varepsilon \ll 1, \tag{11}$$

with $f(\mathbf{y}, t) = \exp(i\mathbf{k} \cdot \mathbf{y} + \lambda t)$, $i = \sqrt{-1}$, and $\mathbf{k} = k_1\mathbf{e}_1 + k_2\mathbf{e}_2$.

In the absence of a perturbation ($\varepsilon = 0$). i.e., when the interface is flat, the corresponding zero-order problem has for solution the fundamental solution, the stability of which is investigated. The infinitesimal scalar $\varepsilon \ll 1$ is responsible for the first-order effect of the perturbation in units of Y. The perturbation has a wavy pattern resulting from the superposition of waves characterized by the polarization vector $\mathbf{k}$ with norm $k = \sqrt{k_1^2 + k_2^2}$

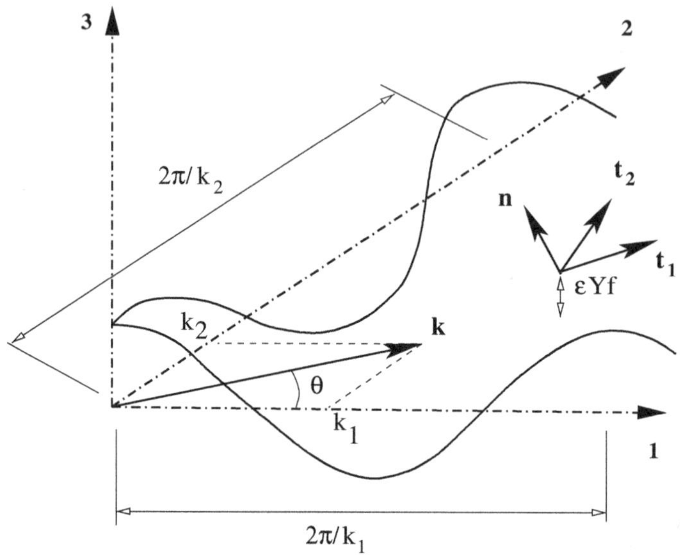

Figure 2. The geometry of the perturbed interface between the solid and the fluid phases

and orientation angle θ with respect to the first coordinate axis (cf. Figure 2). To every mode (k, θ) corresponds a complex scalar λ called the stability exponent which determines its rate of growth or decay with time t. The collection of all stability exponents determines the stability of the fundamental solution. If for given stress conditions, there exists a vector $\mathbf{k}$ associated with a λ with a positive real part, then the stressed system is said to be unstable (exponential growth of the third component of the position vector $\mathbf{y}$). Conversely, if for a given stress state, all the vectors $\mathbf{k}$ have a λ with a negative real part, then the system is stable. The conditions found at the transition between stability and instability define neutral stability. One should recall in this general introduction on stability that in the framework of complex variable developments, the physical part of every parameter or variable is the real part of the relevant expression.

Classical results in the differential geometry of surfaces in three dimensions (see for instance Aris, 1989) are now applied to the perturbed interface. The first step is the definition of a local orthogonal direct basis. Two covariant tangent vectors, not necessarily unitary, may be expressed as

$$\mathbf{t}_\alpha \equiv \frac{\partial \mathbf{y}}{\partial y^\alpha} = \mathbf{e}_\alpha + i\varepsilon Y k_\alpha f(\mathbf{y}, t)\mathbf{e}_3 \,. \tag{12}$$

We recall that Greek subscripts range from one to two whereas Latin indices ranges from one to three. The two tangent vectors are orthogonal to the second order in perturbation, i.e., their dot product is $o(\epsilon^2)$. In order to complete the basis, the normal vector, pointing towards the fluid is calculated as the cross product $\mathbf{t}_1 \times \mathbf{t}_2$, so that the triplet $\{\mathbf{t}_\alpha, \mathbf{n}\}$ is an orthogonal direct basis.

$$\mathbf{n} = \mathbf{e}_3 - i\varepsilon Y k_\alpha \mathbf{e}_\alpha f(\mathbf{y}, t) \,, \tag{13}$$

(with the appropriate summation convention on repeated subscripts).

The mean curvature is the other geometric quantity which appears in the surface boundary conditions. It has to be computed for the perturbed interface. The first fundamental form to be computed is the surface metric which covariant components are

$$a_{\alpha\beta} \equiv \mathbf{t}_\alpha \cdot \mathbf{t}_\beta = \delta_{\alpha\beta} + o(\epsilon^2) \,, \tag{14}$$

in which $\delta_{\alpha\beta}$ is the Kronecker symbol (equal to 1 if $\alpha = \beta$, 0 otherwise). The metric, to second order in perturbation, is the same as the metric for the flat interface. Neglecting these second order terms, the contravariant components of the metric tensor are thus simply $\delta^{\alpha\beta}$, with the same definition of the Kronecker symbol. The second tensor needed to compute the mean curvature is the second fundamental form or the curvature tensor which symmetric components are

$$b_{\alpha\beta} \equiv \frac{\partial^2 \mathbf{y}}{\partial y^\alpha \partial y^\beta} \cdot \mathbf{n} = 0 - \varepsilon k_\alpha k_\beta Y f(\mathbf{y}, t) \,. \tag{15}$$

The curvature tensor is zero for the flat interface and reduces to first-order components in ε for the perturbed geometry. The mean curvature $2H$ is by definition the trace of the curvature tensor obtained by contraction with the contravariant surface metric components

$$2H \equiv a^{\alpha\beta} b_{\alpha\beta} = 0 - \varepsilon k^2 Y f(\mathbf{y}, t) \,. \tag{16}$$

Its first order term is thus proportional to the square of the norm of the polarization vector.

3.2 Perturbation in Other Field Variables

The structure proposed above for the perturbed geometry is now generalized to any field variable $A(\mathbf{x}, t)$. This generic variable is taken as the sum of two terms

$$A(\mathbf{x}, t) = A^0(\mathbf{x}) + \varepsilon \tilde{A}(\mathbf{x}) \exp(\lambda t) . \tag{17}$$

The first term $A^0(\mathbf{x})$ can vary spatially but is time independent since it is associated with the fundamental solution of the problem (perfectly flat interface). The second term, infinitesimal in size, is expressed as the product of a spatial variation and an exponential time dependence for growth or decay, according to the sign of the real part of λ. If the fundamental solution is further assumed to depend only on the third coordinate x_3, then the equation (17) is replaced by

$$A(\mathbf{x}, t) = A^0(x_3) + \varepsilon \tilde{A}(x_3) \exp(i\mathbf{k} \cdot \mathbf{x} + \lambda t) , \tag{18}$$

corresponding to the classical normal mode decomposition. This general structure is now applied to state the equations of equilibrium and boundary conditions for the zero and first-order problems, the latter obtained by linearization.

3.3 Linearization of the Boundary Conditions and Initial Stress States

The solid-fluid interface is the locus of stress compatibility conditions between the two phases which have been established in section 1. These are the Laplace conditions which express the boundary conditions of our problem. The solid may sustain non-hydrostatic stresses whereas the fluid is by definition hydrostatic, with a pressure p_f of constant magnitude in space and time, disregarding any viscous effects. The interface is assumed isotropic. These assumptions lead to the Laplace conditions presented in (3), now written as

$$\begin{aligned} [\![\boldsymbol{\sigma}]\!] : \mathbf{n} \otimes \mathbf{n} &= -2H\gamma , \\ [\![\boldsymbol{\sigma}]\!] : (\mathbf{n} \otimes \mathbf{t}_\alpha + \mathbf{t}_\alpha \otimes \mathbf{n}) &= 0 . \end{aligned} \tag{19}$$

This compact notation will prove more convenient for the linearization. It is the one found for instance in Salençon (2001). The double dots : stand for a scalar product between two second order tensors, as already seen in a previous section. The dyadic product symbol $\otimes$ used between two vectors $\mathbf{a}$ and $\mathbf{b}$ results in a second-order tensor which components are

$a_i b_j$. It is also defined in terms of its action upon an arbitrary vector $\mathbf{c}$ by $\mathbf{a} \otimes \mathbf{b} \cdot \mathbf{c} = \mathbf{a}(\mathbf{b} \cdot \mathbf{c})$.

Making use of the structure (18) for the perturbation in stress and in the normal and tangent vectors, respectively (13) and (12), one obtains, grouping terms of the same order in ε and neglecting terms of order higher than one ($o(\epsilon^2)$ and higher).

$$\begin{aligned} &[\![\boldsymbol{\sigma}^0]\!] : \mathbf{n}^0 \otimes \mathbf{n}^0 + \varepsilon\big([\![\tilde{\boldsymbol{\sigma}}]\!] : \mathbf{n}^0 \otimes \mathbf{n}^0 + [\![\boldsymbol{\sigma}^0]\!] : 2\tilde{\mathbf{n}} \otimes \mathbf{n}^0\big) \\ &\qquad\qquad = -(2H^0 + \varepsilon 2\tilde{H})\gamma\,, \\ &[\![\boldsymbol{\sigma}^0]\!] : (\mathbf{n}^0 \otimes \mathbf{t}^0_\alpha + \mathbf{t}^0_\alpha \otimes \mathbf{n}^0) + \varepsilon\{[\![\tilde{\boldsymbol{\sigma}}]\!] : (\mathbf{n}^0 \otimes \mathbf{t}^0_\alpha + \mathbf{t}^0_\alpha \otimes \mathbf{n}^0) + \\ &\qquad\qquad 2[\![\boldsymbol{\sigma}^0]\!] : (\tilde{\mathbf{n}} \otimes \mathbf{t}^0_\alpha + \tilde{\mathbf{t}}_\alpha \otimes \mathbf{n}^0)\} = 0. \end{aligned} \tag{20}$$

Retaining only the terms independent of ε, the zero-order boundary conditions read

$$\sigma^0_{33} = -p_f\,, \quad \sigma^0_{3\alpha} = 0\,. \tag{21}$$

They express the action of the fluid pressure on the flat interface and conditions of zero out of plane shear component. Only three out of the six stress components are defined by the Laplace conditions.

If one further assumes that the stress state is homogeneous over the whole half space occupied by the solid phase and that the principal stress directions are aligned with the axes of the coordinate system introduced in Figure 2,the expression of the stress tensor in the solid becomes

$$\boldsymbol{\sigma}^0 = -p_f \mathbf{e}_3 \otimes \mathbf{e}_3 + \sum_{\alpha=1}^{2} \sigma^0_\alpha \mathbf{e}_\alpha \otimes \mathbf{e}_\alpha\,, \tag{22}$$

with only two unknowns, the principal stresses σ^0_α, which are independent of the coordinates x_α. This property allows us to use the normal-mode decomposition presented in (18) in the rest of this study.

We now turn our attention back to the system of equations (20) and collect terms proportional to the small parameter ε. The resulting equations

$$\begin{aligned} &[\![\tilde{\boldsymbol{\sigma}}]\!] : \mathbf{n}^0 \otimes \mathbf{n}^0 + [\![\boldsymbol{\sigma}^0]\!] : 2\tilde{\mathbf{n}} \otimes \mathbf{n}^0 = -2\tilde{H}\gamma\,, \\ &[\![\tilde{\boldsymbol{\sigma}}]\!] : (\mathbf{n}^0 \otimes \mathbf{t}^0_\alpha + \mathbf{t}^0_\alpha \otimes \mathbf{n}^0) + 2[\![\boldsymbol{\sigma}^0]\!] : (\tilde{\mathbf{n}} \otimes \mathbf{t}^0_\alpha + \tilde{\mathbf{t}}_\alpha \otimes \mathbf{n}^0) = 0\,, \end{aligned} \tag{23}$$

are the boundary conditions of the first-order problem. Recalling now the fundamental solution of the zero-order problem (equation (22), flat interface of normal $\mathbf{n}^0 = \mathbf{e}_3$ and tangent vectors parallel to the other basis vectors

$\mathbf{t}^0_\alpha = \mathbf{e}_\alpha$) as well as the expression for the first-order perturbation in tangent, normal and curvature (equations (12), (13) and (16) respectively) the system of equations (23) becomes

$$\tilde{\sigma}_{33} = -k2Y\gamma f(\mathbf{y},t)\,, \quad \tilde{\sigma}_{\alpha 3} = k_\alpha Y \bar{\sigma}_\alpha f(\mathbf{y},t) \quad \text{(not summed over } \alpha)\,, \tag{24}$$

where $\bar{\sigma}_\alpha$ denotes the effective principal stress $\sigma^0_\alpha + p_f$, σ_α being the in–plane principal stresses.

Finding the solution to this first-order problem now amounts to the classical search for a three-dimensional stress state that satisfies the boundary conditions expressed in (24) and equilibrium equations (1) in the solid phase. The following subsection deals with an analytical approach to this problem of linear elasticity theory.

3.4 Solution to the First-Order Linear Elasticity Problem

The solution to a well-posed linear elasticity problem is unique and some of these problems can be solved analytically. Among various techniques to construct such analytical solutions to problems in three dimensions, we shall employ an elegant and simple procedure based on the use of Galerkin vectors (cf. Fung, 1965). We first state the linear elasticity problem in general, before turning to the specifics of the perturbation analysis. The field equations of the linear theory of elasticity are obtained by combining Hooke's law

$$\sigma_{ij} = \lambda u_{i,i} + \mu(u_{i,j} + u_{j,i}) \tag{25}$$

and the equilibrium equations (1), resulting in Navier's equations for the displacement vector $\mathbf{u}(\mathbf{x})$

$$\mu u_{i,jj} + (\lambda + \mu) u_{j,ji} = 0. \tag{26}$$

These are written here in the displacement components that appear in the definition of the infinitesimal strain $\epsilon_{ij} = \frac{1}{2}(u_{i,j} + u_{j,i})$, and the Lamé coefficients λ and μ.

A standard means to obtain the solution to linear elasticity problems is to use the method of potentials based on Helmholtz' theorem, according to which any analytic vector field $\mathbf{u}$ can be expressed in the form

$$u_i = \phi_{,i} + e_{ijk}\psi_{k,j}\,, \tag{27}$$

in terms of is a scalar potential ϕ and a vector potential $\boldsymbol{\psi}$ of the vector field $\mathbf{u}$ (the permutation symbol e_{ijk} is equal to +1 for an even permutation of indices, 0 if two indices coincide, and -1 otherwise).

A simple way of introducing Galerkin vectors is now to consider the vector potential itself as being generated by another vector field (for details, see Fung, 1965), which leads one to define the components of the Galerkin vectors as biharmonic functions F_i ($\nabla^4 F_i = 0$, or $F_{i,jjmm} = 0$ in component form), such that $2\mu u_i = 2(1-2\nu)F_{i,jj} - F_{j,ji}$, where ν is Poisson's ratio. In terms of these functions, Hooke's law (25) now becomes

$$\sigma_{ij} = \left[\delta_{ij}\nu\nabla^2 - \frac{\partial^2}{\partial x_i \partial x_j}\right] F_{k,k} + (1-\nu)\nabla^2(F_{i,j} + F_{j,i}) \,, \tag{28}$$

while the equilibrium condition (1) yields the biharmonic equations

$$\nabla^4 F_i = 0. \tag{29}$$

Generalizing Fung's example of a vertical load on the horizontal surface of a semi–infinite solid, we search solutions to (29) of the form1

$$F_i = (A_i + B_i x_3 k)\psi(\mathbf{x}) \,, \tag{30}$$

where A_i and B_i are two constants and ψ an unknown function of the position vector $\mathbf{x}$. The equilibrium condition (29) requires that ψ is a harmonic function

$$\nabla^2 \psi = 0 \,, \tag{31}$$

which can be written as

$$\psi(\mathbf{x}) = \exp(\boldsymbol{\omega}\cdot\mathbf{x}), \quad \text{with } \boldsymbol{\omega} = (ik_1, ik_2, k) \,, \tag{32}$$

to ensure the decay of ψ for x_3 tending to $-\infty$. The equilibrium problem has thus been reduced to the determination the three pairs of scalars (A_i, B_i) from the relevant boundary conditions at the solid-fluid interface. In this elasticity problem, we shall use the principle of superposition and construct the solution by addition, one Galerkin vector F_3 satisfying the boundary condition relative to the normal component of the traction vector and the other two equilibrating the shear components.

Normal component: Expressing (28) for F_3, and stating that the shear components are vanishing, one gets:

$$\begin{aligned} \sigma_{\alpha 3} &= -F_{3,33\alpha} + (1-\nu)\nabla^2 F_{3,\alpha} = 0, \\ \sigma_{33} &= (2-\nu)\nabla^2 F_{3,3} - F_{3,333} \,, \end{aligned} \tag{33}$$

and replacing F_3 according to (30),

$$\begin{aligned} \sigma_{\alpha 3} &= -ik_\alpha k_2 \, (2\nu B_3 + A_3)\, \psi(x_1, x_2, 0) = 0 \,, \\ \sigma_{33} &= k_3 \, ((1-2\nu)B_3 - A_3)\, \psi(x_1, x_2, 0) \,, \end{aligned} \tag{34}$$

leading to

$$\begin{aligned} A_3 &= 2\nu B_3\,, \\ \sigma_{33} &= k_3 B_3 \psi(x_1, x_2, 0)\,. \end{aligned} \tag{35}$$

Shear or tangential conditions: A similar procedure will allow the determination of the two additional vectors F_α. Equations (28) yield the system :

$$\begin{aligned} \sigma_{\alpha 3} &= -F_{\beta,\beta 3\alpha} + (1-\nu)\nabla^2 F_{\alpha,3}, \\ \sigma_{33} &= \nu F_{\beta,\beta\alpha\alpha} - (1-\nu) F_{\beta,\beta 33} = 0\,, \end{aligned} \tag{36}$$

which may again be expressed as,

$$\begin{aligned} \sigma_{\alpha 3} &= \left(k k_\alpha k_\beta (A_\beta + B_\beta) + (1-\nu) 2 B_\alpha k^3\right) \psi(x_1, x_2, 0)\,, \\ \sigma_{33} &= -i k_\beta k_2 \left((1-\nu) 2 B_\beta + A_\beta\right) \psi(x_1, x_2, 0) = 0\,. \end{aligned} \tag{37}$$

The last equation in (37) provides

$$A_\beta = -2(1-\nu) B_\beta\,, \tag{38}$$

which is then replaced in the first two equations (37) in order to yield the expressions for B_β (by resolution of the linear system),

$$B_\beta = \frac{1}{2k^5(1-\nu)} \left[2(1-\nu)\delta_{\alpha\beta} k^2 + (1-2\nu)(k_\alpha k_\beta - k^2 \delta_{\alpha\beta})\right] \sigma_{\alpha 3}\,. \tag{39}$$

The six scalars entering the definition of the three Galerkin vectors are thus computed in term of stresses at the interface and one may go back to equations (28 to get the two stress components required in the linear stability analysis σ_{11} and σ_{22} ,

$$\begin{aligned} \sigma_{11} &= \sigma_{33}(2\nu + (1-2\nu)K_1^2) + 2iK_2\sigma_{13} + 2i\nu K_2^2 K_\alpha \sigma_{3\alpha}\,, \\ \sigma_{22} &= \sigma_{33}(2\nu + (1-2\nu)K_2^2) + 2iK_2\sigma_{23} + 2i\nu K_1^2 K_\alpha \sigma_{3\alpha}\,, \end{aligned} \tag{40}$$

in which K_α stands for k_α/k. We now retrun to our specific problem and, making use of the boundary conditions as expressed by (24), the needed components of the perturbed stress are:

$$\begin{aligned} \tilde{\sigma}_{11} &= \{-k^2\gamma(2\nu + (1-2\nu)K_1^2) \\ &\quad - 2k(K_1^2\bar{\sigma}_1(1+\nu K_2^2) + \nu K_2^4 \bar{\sigma}_2)\} Y f(\mathbf{y}, t)\,, \\ \tilde{\sigma}_{22} &= \{-k^2\gamma(2\nu + (1-2\nu)K_2^2) \\ &\quad - 2k^2 K_2^2 \bar{\sigma}_2(1+\nu K_1^2) + \nu K_1^4 \bar{\sigma}_1)\} Y f(\mathbf{y}, t)\,. \end{aligned} \tag{41}$$

The scalars $K_\alpha = k_\alpha/k$ define the orientation of the polarization vector $\mathbf{k}$. Since θ is the angle between $\mathbf{k}$ and the axis 1, then $K_1 = \cos\theta$ and $K_2 = \sin\theta$ (cf. Fig. 2).

4 Stability Criterion for the Interface Morphology

The linear stability analysis presented in the previous section is applied to study the evolving morphology of the solid/fluid interface. The solid is modeled as a half-space under stress in equilibrium with a fluid solution. The only deformation mechanism is pressure solution. In the second part of this section, we give the proof that the polarization vector is aligned with the direction of maximum compressive principal stress.

4.1 The Stability Exponent

The motion of the solid/fluid interface relative to the solid material is driven by the kinetic law (9). This law is linear(10) in terms of the thermodynamic driving force χ for pressure solution defined in (8). We recall that in the following, the chemical potential of the solute μ_S which enters the definition of χ is assumed constant in space and time. The linearization of the driving force χ is thus the result of the linearization of the Helmholtz free energy of the solid phase (5), the material density (6) and the curvature (16). These quantities are expressed in terms of the perturbation in geometry (presented in section 3.1) and stress (developed in section 3.4). These intermediate steps based on a lengthy yet straightforward computation are not detailed here. The perturbed term of the linearized kinetic law may then be expressed as:

$$\begin{aligned}\tilde{J}_N = L\tilde{\chi} \; = \; & k^2\frac{\gamma Y}{\rho}\{1 - \frac{(1-2\nu)\rho}{E\rho_0}\left[\bar{\sigma}_{11}(\nu + K_1^2 - \nu K_2^2) + \right. \\ & \left. \bar{\sigma}_{22}(\nu - \nu K_1^2 + K_2^2)\right]\} \\ & -k\frac{2Y}{E\rho_0}\{\bar{\sigma}_{11}^2 K_1^2(1+\nu K_2^2 - \nu^2 K_1^2) + \bar{\sigma}_{22}^2 K_2^2(1 + \nu K_1^2 - \nu^2 K_2^2) \\ & -\bar{\sigma}_{11}\bar{\sigma}_{22} 2K_1^2 K_2^2 \nu(1+\nu)\}\,. \end{aligned} \tag{42}$$

The linearization of the mass flux, defined in (7) provides the other expression of $\tilde{J}_N$ as a linear function of λ:

$$\tilde{J}_N = -\lambda\rho_{s0} Y f\,. \tag{43}$$

Equating the right hand sides of (42) and (43), we would get explicitly the stability exponent λ. However, one may wish to consider a simpler expression by accounting for the fact that the magnitudes of elastic stresses are always small compared to Young's modulus and thus terms in $\bar{\sigma}_{ij}/E$ may be disregarded as very small compared to 1. The reduced expression for λ becomes:

$$\begin{aligned}\lambda\rho_{s0}Y &= -k^2\frac{\gamma Y}{\rho} + k\frac{2Y}{E\rho_0}\{\bar{\sigma}_{11}^2K_1^2(1+\nu K_2^2-\nu^2K_1^2) \\ &\quad +\bar{\sigma}_{22}^2K_2^2(1+\nu K_1^2-\nu^2K_2^2) - \bar{\sigma}_{11}\bar{\sigma}_{22}2K_1^2K_2^2\nu(1+\nu)\}\,.\end{aligned} \tag{44}$$

The next step is to normalize this expression by the introduction of dimensionless wavenumbers and stresses. The perturbation unit Y may be equated to the only length scale of the problem, namely γ/E The dimensionless wavenumber is thus $kY = k\gamma/E$ and the stress may be normalized by E. The dimensionless form of (44) is thus :

$$\frac{\lambda\rho_0^2\gamma}{LE^2} = -k^2 + 2kF(\bar{\sigma}_{\alpha\beta}, K_\delta) \tag{45}$$

with

$$\begin{aligned}F(\bar{\sigma}_{\alpha\beta}, K_\delta) = \;& \bar{\sigma}_{11}^2K_1^2(1+\nu K_2^2-\nu^2K_1^2) + \bar{\sigma}_{22}^2K_2^2(1+\nu K_1^2-\nu^2K_2^2) \\ & -\bar{\sigma}_{11}\bar{\sigma}_{22}2K_1^2K_2^2\nu(1+\nu)\,,\end{aligned} \tag{46}$$

where k and $\bar{\sigma}_{\alpha,\beta}$ are the dimensionless quantities defined above, so that a dimensionless stability exponent may also be defined in a straightforward way as:

$$\Lambda \equiv \frac{\lambda\rho_0^2\gamma}{LE^2} \tag{47}$$

Note that the stability exponent is a real number so that no flutter-type of instabilities (oscillatory in time) may occur in our problem.

4.2 Stability Analysis

It has been recalled in section 3.1 that the stable or unstable character of the solution rested on the sign of the real part of a stability exponent. The analysis of stability is therefore based on the sign of Λ in the dimensionless equation

$$\Lambda = -k^2 + 2kF(\bar{\sigma}_{\alpha\beta}, K_\delta)\,, \tag{48}$$

giving Λ as a sum of two terms. The first is always negative and proportional to the square of the wave number; it is related to the surface tension. The second term is a function of the effective stress and proportional to the wavenumber. We shall establish in the following that it is always positive for the dominant perturbation. In analogy with 2D stability analysis, we show that the short wavelengths are always stable because of the dominant role of surface tension and that the destabilizing factor is the effective stress, the difference between the solid stress state and the fluid pressure.

Let us now turn to the dominant mode, defined as the mode having the maximum rate of growth, it may be characterized by two scalars, the dimensionless wavenumber k_M and the orientation of the polarization vector in the (1-2) plane defined by the angle θ_M (recall the last remark of section 3 and Figure 2). The dominant perturbation is thus characterized by two differential equations expressing its extremal property.

$$\Lambda_M(k_M, \theta_M) \quad \text{such that} \quad \frac{\partial \Lambda}{\partial k} = 0\,, \quad \frac{\partial \Lambda}{\partial \theta} = 0\,. \tag{49}$$

The first partial derivative in (49) is computed from (48) and is set to zero to provide the dominant dimensionless wavenumber

$$k_M = F(\bar{\sigma}_{\alpha\beta}, \theta_M)\,, \tag{50}$$

in terms of the effective stresses and the angle θ_M. This angle is determined from the second condition in (49) which, from (48), reads

$$\frac{\partial F}{\partial \theta}(\bar{\sigma}_{\alpha\beta}, \theta) = 0 \tag{51}$$

The left hand side of this equation may be expressed as a quadratic form in stress components $\bar{\sigma}_{11}, \bar{\sigma}_{22}$,

$$\frac{\partial F}{\partial \theta}(\bar{\sigma}_{\alpha\beta}, \theta) \equiv 2K_1K_2\,\{\bar{\sigma}_{11}, \bar{\sigma}_{22}\}\,[A(\theta)] \left\{ \begin{array}{c} \bar{\sigma}_{11} \\ \bar{\sigma}_{22} \end{array} \right\} \tag{52}$$

with the symmetric array $[A(\theta)]$,

$$[A(\theta)] = \left[\begin{array}{cc} -1 - \nu K_2^2 + \nu(1+2\nu)K_1^2 & -\nu(1+\nu)(K_1^2 - K_2^2) \\ -\nu(1+\nu)(K_1^2 - K_2^2) & 1 + \nu K_1^2 - \nu(1+2\nu)K_2^2 \end{array} \right],$$

Equation (51) is a product of three terms : it has two obvious solutions $K_1 = 0$ and $K_2 = 0$ and possibly a third one if the quadratic form has a real root. The determinant of the array $A(\theta)$ is $-(1-\nu^2)^2$, a negative scalar, independent of θ. Consequently, there is no non zero stress state for which this third term is zero. The only solutions are thus,

$$\theta_M = 0 \quad (\text{mod } \pi) \quad \text{and} \quad \theta_M = \pi/2 \quad (\text{mod } \pi)\,. \tag{53}$$

These two values of θ_M provide extrema for F, in order to find the maximum, one must study the sign of

$$\frac{\partial^2 F}{\partial \theta^2}(\bar{\sigma}_{\alpha\beta}, \theta)\,, \tag{54}$$

which is the sign of the expression

$$(K_1^2 - K_2^2)\,\{\bar{\sigma}_{11}, \bar{\sigma}_{22}\}\,[A(\theta)]\left\{\begin{matrix}\bar{\sigma}_{11}\\ \bar{\sigma}_{22}\end{matrix}\right\} + K_1 K_2 \frac{\partial}{\partial \theta}\left(\{\bar{\sigma}_{11}, \bar{\sigma}_{22}\}\,[A(\theta)]\left\{\begin{matrix}\bar{\sigma}_{11}\\ \bar{\sigma}_{22}\end{matrix}\right\}\right)\,, \tag{55}$$

for the two values of θ given in (53). This translates into two expressions

$$-(1+\nu)(\bar{\sigma}_{11} - \bar{\sigma}_{22})((\bar{\sigma}_{11} + \bar{\sigma}_{22}) - 2\nu\bar{\sigma}_{11}) \text{ for } \theta_M = 0\,, \tag{56}$$

and

$$-(1+\nu)(\bar{\sigma}_{22} - \bar{\sigma}_{11})((\bar{\sigma}_{11} + \bar{\sigma}_{22}) - 2\nu\bar{\sigma}_{22}) \text{ for } \theta_M = \pi/2. \tag{57}$$

Let us assume that the material is brittle in tension, so that the discussion of the instability is limited to negative (or compressive) principal effective stresses. Let $\bar{\sigma}_{22}$ be the major compressive stress, one thus considers only stress states such that

$$\bar{\sigma}_{22} \leq \bar{\sigma}_{11} \leq 0\,. \tag{58}$$

Under this strong assumption, one may then assess that the expression (57) is negative and (56) is positive. The only value to keep for θ_M is thus $\pi/2$, which means that the dominant mode has its normal parallel to the major compressive stress direction.

The dominant instability is always a wave which front is perpendicular to the direction of the effective stress of larger magnitude. The wavenumber of the dominant mode is

$$k_M = F(\sigma_{\alpha\beta}, \theta_M = \pi/2) = (1-\nu^2)\bar{\sigma}_{22}^2 \tag{59}$$

and its rate of growth $\Lambda_M = k_M^2$. Neutral stability conditions are defined by $\Lambda = 0$ and are obtained for a wavelength $k = 2F$.

5 Application to the Experiments of den Brok and Morel

The stability criterion, obtained for a homogeneous stress state over an infinite half space, is now applied as a local criterion to the experiments conducted by den Brok and Morel (2001). The first part of this section is a concise description of their set-up and main results. The second part presents our stability predictions which requires as input a local knowledge of the stress tensor. In order to determine to stress state at any point of the surface of the perforated plate used by den Brok and Morel, we have made a finite element computation, with the assumption of linear elasticity and plane stress conditions.

5.1 Description of the Experiments

The experimental set-up chosen by Morel and den Brok (2001) consisted of a vessel in which an aqueous phase (the solvent) is brought into contact with two solid specimens at atmospheric pressure. The well stirred solution was maintained under-saturated so as to keep the dissolution process going. The solid specimens were made of single K-alum crystals of cubic symmetry, a highly soluble salt; they had the shape of a rectangular plate, measuring $4 \times 6 \times 10$ mm. A cylindrical hole, 2 mm in diameter, was drilled through these plates, which allowed the creation of a stress gradient when a plate was compressed perpendicular to its smallest (4×6 mm) face. At room temperature, the K-alum salt is said to respond in an elastic-brittle way to loading, with a uniaxial compressive strength of 20 MPa.

One of the two immersed specimens is loaded up to a few MPa. The other is exposed solely to the fluid pressure and is used for comparison purposes; this specimen develops a surface waviness that is described as an anastomosing pattern of fine etch grooves with a wavelength of the order of one μm. The compressed specimen also developed a pattern of etch grooves on its surface, but these were oriented perpendicular to the direction of the largest principal stress. Also, close to the hole, the grooves are bent towards a radial direction at its circumference. The observed wavelengths were of the order of 20 μm and tended to decrease with larger loads. A few hours after the removal of the load this marked pattern changed into the anastomosing structure of the unloaded comparison sample. This reversible trend strongly suggests that the observed changes in surface morphology were not related to dislocation structures, but arose from load-induced changes in the stored elastic energy of the specimen.

The observed etch grooves do not have the shape of the smooth trigonometric function used to construct the normal modes of our stability analysis. The dissolved regions do resemble a sine wave, but there are no bulging regions requiring deposition. This difference is illustrated in Figure 3 in cross section.

The trigonometric function and the observed groove are displayed as dotted and solid curve, respectively. The two curves are coincident in the dissolution region. The absence of deposition during the development of the

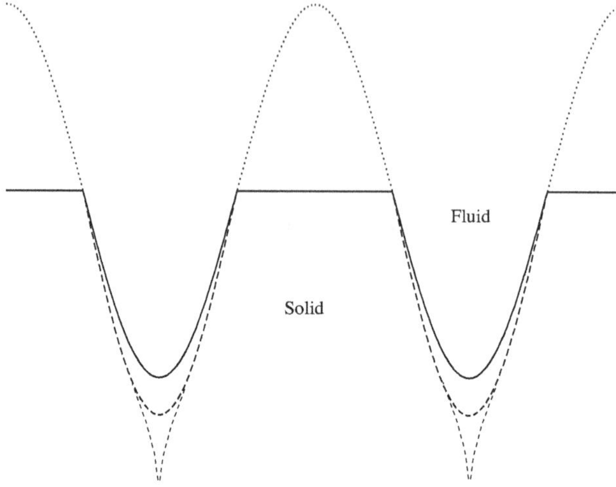

Figure 3. The shape of the wavy pattern observed in the experiments of den Brok and Morel (solid curves) corresponds approximately to the trigonometric perturbation (dotted curves) introduced in the linear stability analysis, if precipitation is suppressed in the latter. A non-linear stability analysis (Ghoussoub and Leroy, 2001) shows that the wavy pattern evolves gradually from the smooth perturbation to form cusps where the stress concentration and the dissolution rates are largest.

instability is very likely due to the difference in characteristic times associated with the kinetics of dissolution and precipitation (Lehner and Leroy, 2004). This difference is not modeled by the simple, linear kinetic law used in our analysis. In spite of this, we shall nevertheless follow the reasoning of den Brok and Morel in the next subsection and, using the proposed linear stability criterion as a local condition, analyze their experimental observations.

5.2 Comparison with the Predictions of Linear Stability Theory

The stress gradient over the loaded specimen is most conveniently computed by use of a finite-element approximation. Four-noded elements are selected for this purpose, with four quadrature points. We assume a plane state of stress, given that the out-of-plane stress is of the order of the (atmospheric) fluid pressure and can thus be neglected against the stress induced by the applied load (nominal value of 5 MPa). The value of Young's modulus, which appears in the normalization of stresses magnitudes is taken as 20 GPa.

The computed results are presented as four graphs in Figure 4, each covering only one quarter of the plate for reasons of symmetry. In the the first plot, Figure 4a, the dimensionless vertical stress σ_{22} is slightly positive (maximum value 0.13×10^{-4} or 0.26 MPa) in a small region near the top of the hole and then reaches as expected a maximum compressive value (-7.0×10^{-4} or 14 MPa) near the horizontal diameter, which corresponds to a stress amplification of about three of the average compression applied at the top of the plate. Den Brok and Morel (2001) did observe a small damaged region at the top of the hole, which may coincide with our area of computed positive vertical stress, showing the brittle behavior of the specimen under tensile forces. Owing to the small region involved and the small positive magnitude of the stress, we do not recompute the stress field over the specimen, but limit the discussion to the regions of well-established compressive regime. The vertical direction does not coincide everywhere with the major principal stress direction, as shown on Figure 4b. This direction is indicated by an arrow, with isocontours every 5 degrees.The most drastic changes of principal direction are around the traction free hole, where the stress field differs most from a simple uniaxial compression.

The results shown in Figure 4b compare well with the orientation of the wavy pattern observed by den Brook and Morel (their Figure 3b). The (in-plane) normal direction of the observed grooves to the left of the hole is indeed approximately 95°, as found by our computation. This direction changes quickly close to the hole and become radial at its circumference, which is consistent with a theoretically expected direction orthogonal to the maximum compressive stress direction.

Next we compare the wavelengths of the surface instability with that of the computed dominant mode, assuming a surface tension of 5×10^{-2} Pa m and again a value of Young's Modulus of 20 GPa. The wavelength distribution is presented in Figure 4c with isocontours ranging from 0 to 1000 μm and more. This range is selected so as to bring out the distribution to the left of the hole, where a comparison with the observations is possible.

The predicted wave lengths of the order of 250 μm match the observations in that region. However, very large wavelengths of 9000 μm are computed above the hole and are associated with a low stress region. They are not shown specifically on this graph, as they belong to the zone $> 1000 \mu m$, and they have not been observed in the experiments. The smallest computed wavelengths (50 μm and less) are close to the hole in the region of high compressive stress, they have also not been observed by den Brok and Morel.

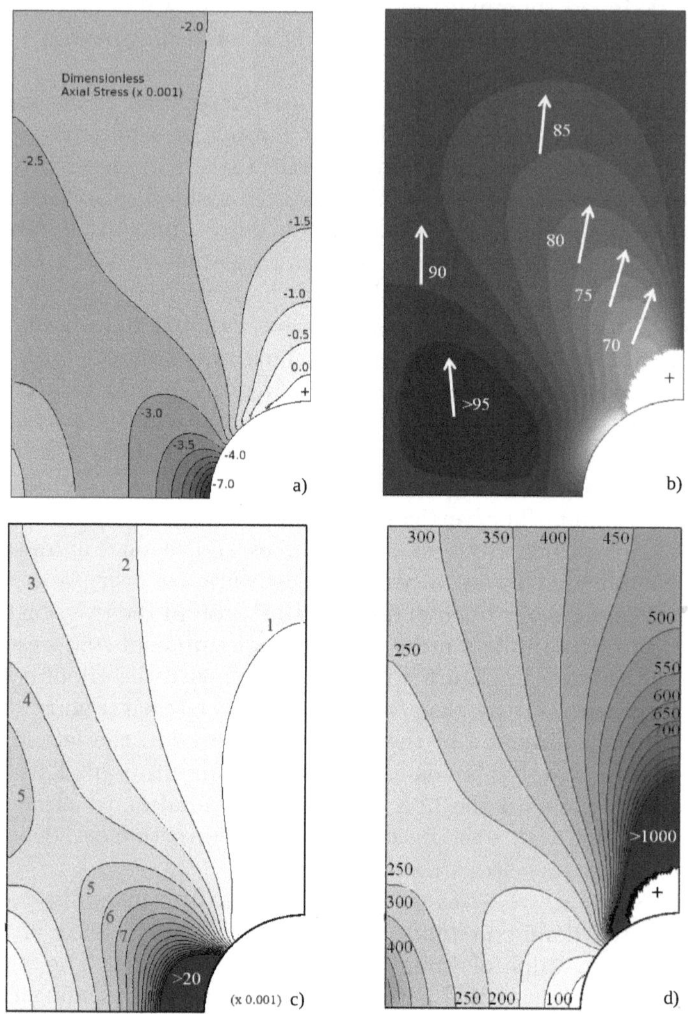

Figure 4. a) Isocontours of dimensionless vertical stress σ_{22}, ranging from -7.0×10^{-4} to 0. with a maximum of 0.13 in the area of positive values; b) Isocontours of the principal compressive stress direction, ranging from $\leq 60°$ to $95°$; c) Isocontours of the wavelength of the instability, ranging from 0 to ≥ 1000 μm ; d) 20 isocontours of the dimensionless rate of growth of the dominant mode of instability, ranging from 0 to 0.01.

This insensitivity of the wavelength to the magnitude of the vertical compressive stress has been mentioned by den Brok and Morel (2001). In our model, its theoretical expression is nevertheless very sensitive to the stress magnitude since the wavelength varies like $1/\sigma_{22}^2$. In the model, the rate of growth of the dominant mode of instability is also very much dependent on the major principal stress (varying like σ_{22}^4, so that one would expect much more pronounced ridges in the higher stressed regions. Figure 4d presents the distribution of the dimensionless rate of growth of instability in the form of isocontours ranging from 0 to 0.02 . The scaling is based on the kinetics of quartz, considered by Ghoussoub and Leroy (2001), in the absence of any information on the kinetics for K-alum. However, what is important is the variation of this growth rate, i.e., the stability exponent as defined in (47) over the specimen, not its absolute value. The upper bound to the range plotted is selected so as to allow an observation of the gradient over the entire domain; note, however, that isocontours have not been plotted in a region close to the hole where a very sharp increase of the rate of growth of the instability is computed. Indeed a stress concentration of 3 may lead to 2 orders of magnitude on the rate of growth of instability. Such a local variation in the rate of growth of the instability, or rate of dissolution, should result in grooves with very different amplitude which could not be missed in the laboratory. This feature of the surface morphology was not reported by den Brok and Morel (2001). However, we lack the information of how the wavy pattern developed with time. It is likely that the initial instability appeared near the most stressed region, but then stopped when the departure from a plane surface reached a certain maximum value, relaxing or altering sufficiently the local state of stress.

There remain therefore significant discrepancies between the predicted morphology evolution and the observations, so that one must conclude that the physics necessary for modeling the observations may be more complex than assumed here. In particular, the temporal evolutions of the surface geometry are not accounted for. But, on the other end, we have only access to the initial and final observation of the surface.

6 Concluding Remarks

An instability may affect the surface of a stressed elastic solid when it is brought in contact with a solvent. A theory of pressure solution has been applied in this chapter to derive a stability criterion based on a linear analysis that uses a normal mode decomposition. It has been found that the predicted surface instability is dominated by a mode that has its polarization vector parallel to the maximum principal stress. The surface instability, as a pattern of grooves over the initially flat surface, is thus a marker of the local stress state.

A surface instability observed experimentally under comparable conditions by den Brok and Morel has been interpreted in terms of this linear stability theory. This comparison reveals that the observed wavy pattern was indeed aligned consistently with the dominant instability mode over the stressed surface, even though the assumed in-plane isotropy was inappropriate for a crystal. However, the theoretically expected dependence of the wavelength and the time-rate of growth of the dominant mode on stress were not confirmed by the experimental observations. This suggests a morphology evolution that is governed by more complex physics. The history of the development of instability was not recoverable from the experimental data which presented only an initial and a final state. It may well be that our assumptions, valid as the start of the phenomenon of dissolution linked to a local stress gradient became obsolete as the surface geometry and stress state evolved; these evolutions being faster or more drastic in the regions of higher stresses and stress gradients.

Non-linear effects in time might be involved in this, such as have been studied for plane-strain conditions by Ghoussoub (2000) (see also Ghoussoub and Leroy, 2001). Their numerical simulations have shown that a wavy pattern growing with time is accompanied by a stress concentration at the bottom of the grooves (see Figure 3 for a sketch of the predictions of different models). The stress concentration in the region of high curvature increases the driving force for dissolution, despite the stabilizing influence of surface tension. The subsequent evolution should thus be characterized by the formation of two cusps, depicted as thin dashed curves. An observation of the free surface around the hole, perpendicular to the surface of the plate, could have brought some information, but is unfortunately not available. New experiments, preferably at confining fluid pressures high enough to maintain a compressive state of stress in the whole specimen, should therefore be performed in order to shed light on the physics of this complex surface morphology evolution and the complete history of the development of instability should be recorded. Nevertheless, our analysis has shown that in the elastic range, at low stress levels but in the presence of a stress gradient, surface roughening by the development of a surface instability is a dominant mode activated by pressure solution.

Bibliography

Alexander, J.I.D. and W.C. Johnson (1985). Thermomechanical equilibrium in solid-fluid systems with curved interfaces. *J. Appl. Physics* 58, 81–824.

Aris, R. (1989). *Vectors, Tensors and the Basic Equations of Fluid Mechanics.* Dover Publications, Inc. New-York.

Bowen, R.M. (1976). Theory of mixtures, in: *Continuum Physics.* Vol. 3, A.C. Eringen (ed.), Acad. Press: New York, NY.

Bowen, R.M., and J.C. Wiese (1969). Diffusion in mixtures of elastic materials. *Int. J. Engng. Sci.* 7, 689–722.

Coble, R.L. (1963). A model for grain boundary diffusion controlled creep in polycrystalline materials. *J. Appl. Physics* 34, 1679–1682.

De Groot, S.R. and P. Mazur (1962). *Non-equilibrium Thermodynamics.* North Holland: Amsterdam.

den Brok, S.W.J. and J. Morel (2001). The effect of elastic strain on the microstructure of free surfaces of stressed minerals in contact with an aqueous solution. *Geophysical Research Letters* 28, 603–606.

Fung, Y.C. (1965). *Foundations of Solid Mechanics.* Prentice Hall, Englewood Cliffs, NJ.

Ghoussoub, J. (2000). Solid-fluid phase transformation within grain boundaries during compaction by pressure solution. Doctoral thesis, Ecole Nationale des Ponts & Chaussées, Ecole Polytechnique, Palaiseau, France.

Ghoussoub, J. and Y.M. Leroy (2001). Morphology evolution of solid-fluid interfaces within grain boundaries of porous rock. *J. Mech. Phys. Solids* 49, 2385-2430.

Gibbs, J.W. (1878). On the equilibrium of heterogeneous substances. *Trans. Connecticut Academy*, III, pp. 343–524 (May 1877 – July 1878). Reprinted in: *The Scientific Papers of J. Willard Gibbs, Vol. I - Thermodynamics*, Longmans, Green, and Co., Toronto (1906); Dover Edition, New York (1961).

Grinfeld, M. (1982). Phase transitions of the first kind in nonlinear elastic materials. *Mechanics of Solids*, 17, 92–101, (Engl. transl. of *Izv. AN SSSR Mekhanika Tverdogo Tela* 17, 99–109, 1982).

Grinfeld, M. (1991). *Thermodynamic Methods in the Theory of Heterogeneous Systems.* Longmans, Green and Co., Toronto.

Heidug, W.K. (1991). A thermodynamic analysis of the conditions of equilibrium at nonhydrostatically stressed and curved solid-fluid interfaces. *J. Geophys. Res.* 96, 21,909–21,921.

Heidug, W.K. and F.K. Lehner (1985). Thermodynmics of coherent phase transformations in nonhydrostatically stressed solids. *Pure and Applied Geophysics* 123, 91–98.

Heidug, W.K. and Y.M. Leroy (1994). Geometrical evolution of stressed and curved solid-fluid phase boundaries, Transformation kinetics. *J. Geophys. Res.* 99, 505–515.

Hickman, S.H. and B. Evans (1995). Kinetics of pressure solution at halite silica interfaces and intergranular clay films. *J. Geophys. Res.* 100, 13,113–13,132.

Houseknecht, D.W. (1987). Assessing the relative importance of compaction processes and cementation to reduction of porosity in sandstones. *AAPG Bulletin* 71, 633–642.

Houseknecht, D.W. and Hathon, L.A. (1987). Relationships among thermal maturity, sandstone diagenesis and reservoir quality in Pennsylvania strata of the Arkoma basin. *AAPG Bulletin* 71, 568–569.

Lehner, F. K. (1995). A model for intergranular pressure solution in open systems. *Tectonophysics* 245, 153–170.

Lehner, F.K. (1997). Theory of pressure solution creep in wet compacting sediments. In: *Mechanics of Solids with Phase Transformations*, edited by M. Berveiller and F.D. Fischer, CISM Courses and Lectures No. 368, pp. 239–258, Springer-Verlag, Wien - New York.

Lehner, F. K. and J. Bataille (1984/85). Nonequilibrium thermodynamics of pressure solution. *Pure and Applied Geophysics* 122, 53–85.

Lehner, F. K. and Y.M. Leroy (2004). Sandstone compaction by intergranular pressure solution. In: *Mechanics of fluid saturated rocks*, edited by Y. Guéguen and M. Boutéca, Chap.3, pp. 115–168, Elsevier Academic Press, Amsterdam etc.

Leroy, Y.M. and W.K. Heidug (1994). Geometrical evolution of stressed and curved solid-fluid phase boundaries, Stability of cylindrical pores. *J. Geophys. Res.* 99, 517–530.

Machlin, E.S. (1953). Some applications of the thermodynamic theory of irreversible processes to physical metallurgy. *J. of Metals Tans. AIME* 437–445.

Morel, J. and S.W.J. den Brok (2001). Increase in dissolution rate of sodium chlorate induced by elastic strain. *Journal of Crystal Growth* 222, 637–644.

Raj, R. (1982). Creep in polycrystalline aggregates by matter transport through a liquid phase. *J. Geophys. Res.* 87, 4731–4739.

Rimstidt, J.D. and H.L. Barnes (1980). The kinetics of silica-water reactions. *Geochimica et Cosmochimica Acta* 44, 1683–1699.

Rutter, E.H. (1976). The kinetics of rock deformation by pressure solution. *Phil. Trans. R. Soc. London* A 283, 203–219.

Rutter, E.H. (1983). Pressure solution in nature, theory and experiment. *J. Geol. Soc. London* 140, 725–740.

Salençon, J. (2001). *Handbook of Continuum Mechanics: General Concepts, Thermoelasticity*, Springer.

Schutjens, P.M. and C.J. Spiers (1999). Intergranular pressure solution in NaCl: Grain-to-grain contact experiments under the optical microscope. *Oil and Gas Science and Technology-Rev. IFP* 54, 729–750.

Sleep, N.H. (1995). Ductile creep, compaction and rate and state dependent friction within major faults. *Nature* 359, 687–692.

Srolovitz, D.J. (1989). On the stability of surfaces of stressed solids. *Acta Metallurgica* 37, 621–625.

Spiers, C.J. and P.M. Schutjens (1990). Densification of crystalline aggregates by fluid-phase diffusional creep. In: *Deformation Processes in Minerals, Ceramics and Rocks*, edited by D.J. Barber and P.G. Meredith, Chap. 12, pp. 334–353, Unwin Hyman, London.

Spiers, C.J., Schutjens, P.M., Brzesowsky, R.H., Peach, C.J., Liezenberg, J.L. and H.J. Zwart (1990). Experimenmtal determination of constitutive parameters governing creep of rock salt by pressure solution. In: *Deformation Mechanisms, Rheology and Tectonics*, edited by R.J. Knipe and E.H. Rutter, Spec. Publ. No. 54, pp. 215–227, The Geological Society, London.

Truskinovskiy, L.M. (1984). The chemical-potential tensor. *Geochemistry International* 21, 22–36, (Transl. from *Geokhimiya*, No. 12, 1730–44, 1983).

Urai, J.L., Spiers, C.J. Zwart, H.J. and G.S. Lister (1986). Weakening of rocksalt by water during long term creep. *Nature* 324, 554–557.

Weyl, P.K. (1959). Pressure solution and the Force of Crystallization–A Phenomenological Theory. *J. Geophys. Res.* 64, 2001–25.

Introduction to the Finite-Element Method for Elastic and Elasto-Plastic Solids

Yves M. Leroy

Laboratoire de Géologie, CNRS, Ecole Normale Supérieure, Paris, France.

Abstract This introduction to the finite-element method is offered to Earth Scientists with an interest in numerical methods, continuum mechanics and the theory of plasticity. No previous exposure to this material is required. Starting with the theorem of minimum potential energy for a one dimensional poro-elastic column, the finite-element method is introduced as the minimization in the subspace of the set of admissible displacements constructed by local, linear interpolation. The method is then extended to two dimensions to introduce the simplest element, the three-noded triangle. These two examples are also used to introduce the concept of assembly for static problems. The numerical quadrature, and in particular Gauss' and Lobatto's quadrature rules are then presented before constructing the family of Lagrange elements. The finite-element method is then explored in the context of non-linear material response and, in particular, for plasticity, first under the small-strain approximation and then for finite transformations. Examples are presented showing the good performance of the so-called spectral elements in problems involving near incompressible plastic flow, strain localization and buckling. Two exercises and a problem are finally proposed, the latter providing an introduction to these spectral elements which have become popular in earthquake modeling.

1 Introduction

The finite-element (FE) method is now a well established technique for solving complex boundary value problems. The continuum unknowns, such as the displacements or the temperature, are replaced by a set of nodal unknowns thanks to the introduction of a spatial interpolation. A weak form (integral) of the governing equations or the minimization of an energy provides the set of linear or non-linear equations that are satisfied by these nodal unknowns.

The power of modern computers and the availability of commercial codes, such as Abaqus and Nastran or of more academic codes, such as Cesar (LCPC) and Castem (CEA) in France, render possible the rapid construction of the solutions to coupled non-linear problems. The teaching of the finite-element method reflects this potential and has evolved over the last twenty years. The objectives early on were to train developers which could "dive" into finite-element codes and implement new classes of elements, solvers and constitutive relations. The teaching is now geared towards researchers who want to make the best use of the above codes, which required hundred of man-years of development. This chapter is certainly part of the second trend since no code development is proposed. It emphasizes, however, the fundamentals of the method for elastic and elasto-plastic materials, from the algorithmic point of view. It is hoped that this chapter will help the researcher in making educated choices while using the standard finite-element packages.

There is an impressive number of books and papers which present the finite-element method and the following lines do no pretend to contain an exhaustive list but only reflect the author's own experience, as a student, as a researcher and then as a teacher in France and in the USA. The first book on the finite-element method I came across is certainly most appropriate for engineers and is due to Zienkiewicz (1977). The book by Hughes (1987) will complement this first reading. The book by Becker (1981) was instrumental in the preparation of a series of lectures. The finite-element method has attracted the attention of applied mathematicians and the book by Reddy (1991) is an introduction to functional analysis readable by engineers. The concise contribution of Johnson (1987) is an interesting second reading in this direction and the book by Ciarlet (1978) for elliptic problems (typically linear elasticity) is of an advanced level. A very original look on the finite-element method and its reliability is found in the work of Babuśka and Strouboulis (2001).

This chapter also introduces many concepts of continuum mechanics and of the theory of plasticity. Although it has been written to be self-contained, the reader interested to brush up his understanding of continuum mechanics should read the concise book by Chadwick (1976). A more detailed presentation is proposed by Malvern (1969). The presentation of continuum mechanics in this contribution, including the presentation of the finite-element method for linearly elastic solids, is strongly influenced by the teaching of J. Salençon at Ecole Polytechnique (Salençon, 2001; in English). The notation considered in this chapter is close to the one introduced in this last reference.

The book by Hill (1950) remains fundamental to comprehend the theory of plasticity while the contribution of Hughes and Simo (1998) provides the link to computational algorithms. The book by Belytschko et al. (2000) covers both the introduction to the FE method, continuum mechanics and algorithms for finite deformation and plasticity.

This chapter was not presented as a lecture at the CISM course in 2002. Its is proposed here to complement the other contributions which are more theoretical and geared towards analytical solutions, fundamental to understand the various physical problems considered. The numerical methods are required to search beyond the domains of application of these analytical solutions. It should however always be kept in mind that the numerical solutions, provided by the FE method in our particular case, should only be sought once one has developed some reasonable understanding of what the solution should be !

2 The First Example: 1D Compaction for a Poro-Elastic Column

Rather than giving a thorough theoretical development, we prefer a hands-on approach and propose a first example as an introduction to the FE method in 1D. The model problem consists of a poro-elastic column under its own weight and its exact solution is first constructed with the method of displacements. The theorem of minimum potential energy is then presented before discussing the finite-element approximation based on Ritz's method. This first example is also useful for defining various concepts and notations which are used through out this chapter.

2.1 Geometry, Loading, Constitutive Response and Equilibrium Solution

The geometry of the first example consists of a layer of infinite lateral extent and height H sustaining on its top the burden of a fluid layer of thickness e, and clamped on a rigid foundation (Figure 1). The infinite lateral extent implies that the horizontal deformation of the solid layer is impossible and the problem is equivalent to a 1D column of arbitrary cross-section (its surface area is set to one), as illustrated in Figure 1. The column, occupies the domain Ω, and is composed of a fluid-saturated, isotropic rock which constitutive relations are described by the theory of poro-elasticity with undrained conditions. The initial stresses are disregarded. It is assumed that the transformation resulting from the compaction of the column respects the conditions of infinitesimal perturbations corresponding to infinitesimal deformations, rotations and negligible change in the geometry

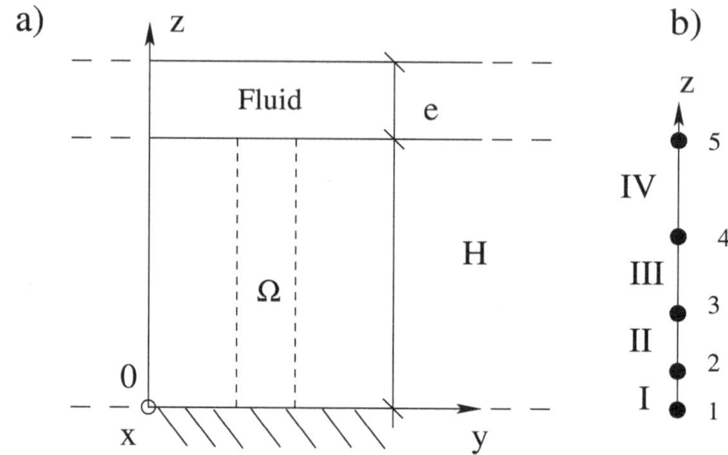

Figure 1. The geometry of the layered structure considered in example 1 is presented in a) and the four elements for the construction of the finite-element approximation in b).

of the deformed column.

The exact solution to this problem is found by the method of displacements which is presented in Figure 2. The starting point consists in postulating the displacement field $\underline{u}(\underline{x})$, function of the material point position $\underline{x}$ in Ω. In our problem, it is assumed that the displacement field has a single degree of freedom, an unknown scalar function of the vertical coordinate. The displacement is thus $\underline{u}(\underline{x}) = u(z)\underline{e}_z$ in which $\{\underline{e}_x, \underline{e}_y, \underline{e}_z\}$ is an orthonormal basis. The origin of the coordinates is set at the contact of the foundation. Note that vectors have been identified with a bar underneath the name of the variable. The unknown function $u(z)$ is continuous, piecewise differentiable and takes a zero value at the origin, on account of the boundary conditions at the base. Any function $\hat{u}(z)$ having these properties is a kinematically admissible displacement field (K.A.) and will be identified by the superposed hat. The set of K.A. displacement field is noted generally as

$$\mathcal{C}(\partial\Omega_{u_i}, u_i^d) = \Big\{\underline{\hat{u}} \,|\, \text{continuous, piecewise differentiable}, \quad \hat{u}_i = u_i^d \ \text{on}\, \partial\Omega_{u_i}\Big\} \tag{1}$$

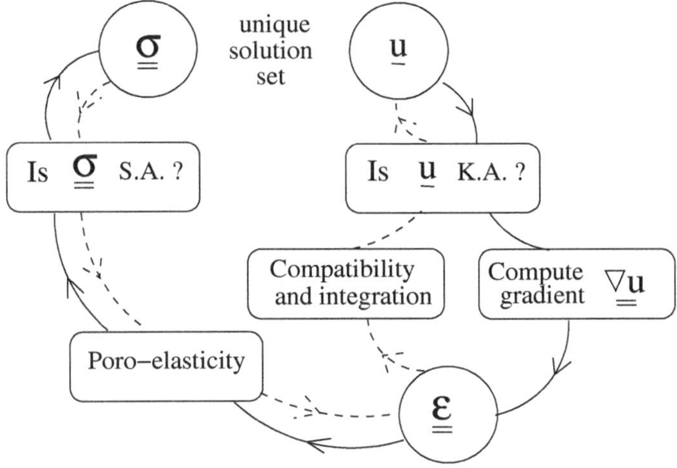

Figure 2. The method of displacements (solid curves) and the method of stresses (dashed curves) to solve boundary-value problems in poro-elasticity. There is a unique set of displacement and stress (kinematically-admissible and statically-admissible, respectively) which are related by the constitutive relations of poro-elasticity.

and depends on the selection of the domain boundary $\partial\Omega_{u_i}$ where some of the displacement components are equal to known values u_i^d (the upper script d stands for "data", known for the specific problem). The first step of the method of displacements ends by making sure that the selected displacement is indeed K.A.

The gradient of the K.A. displacement is $\underline{\underline{\nabla \hat{u}}} = \hat{u}_{,z}\underline{e}_z \otimes \underline{e}_z$ and, since its is symmetric, it is identical to the linearised strain defined by

$$\underline{\underline{\epsilon}}(\underline{\hat{u}}) \equiv \frac{1}{2}(\underline{\underline{\nabla \hat{u}}} + {}^t\underline{\underline{\nabla \hat{u}}}) \,, \quad \left(\epsilon_{ij} \equiv \frac{1}{2}(\frac{\partial u_i}{\partial x_j} + \frac{\partial u_j}{\partial x_i})\right) . \tag{2}$$

The two equations found in (2) are equivalent and the second between large parentheses is proposed to clarify the tensorial notation adopted in the first. Such repetition will be done in what follows for sake of clarify whenever it is found necessary. The linearised strain reads

$$\underline{\underline{\hat{\epsilon}}} = \hat{u}_{,z}\underline{e}_z \otimes \underline{e}_z \,, \tag{3}$$

for our particular case, denoting $\hat{u}_{,z}$ the partial derivative $\partial\hat{u}/\partial z$. Note that in (2) and in what follows, second-order tensors are underlined with a double bar. Note also in (3), the dyadic product $\otimes$. When applied between two vectors $\underline{a} \otimes \underline{b}$, it results in the linear vectorial application which associates to any vector $\underline{v}$ the vector $\underline{a}(\underline{b} \cdot \underline{v})$ in which $\underline{b} \cdot \underline{v}$ is the scalar product between the two vectors $\underline{b}$ and $\underline{v}$. The same definition with an indicial notation is: to v_i is associated the vector $a_i b_j v_j$. The repetition of the indices in the product $b_j v_j$ implies the summation over j from 1 to 3 (or x to z):

$$\underline{b} \cdot \underline{v} = b_j v_j = b_1 v_1 + b_2 v_2 + b_3 v_3 \,.$$

The stress changes due to the straining in (3), in the absence of initial stress, are found from the linear elastic constitutive relations

$$\underline{\underline{\sigma}} = \mathbb{D}^e : \underline{\underline{\hat{\epsilon}}} \quad \text{with} \quad \mathbb{D}^e = (K_u - \frac{2}{3}G)\underline{\underline{\delta}} \otimes \underline{\underline{\delta}} + 2G\, \mathbb{I}_S \,, \tag{4}$$

$$\left(\sigma_{ij} = \mathbb{D}^e_{ijkl} \hat{\epsilon}_{lk} \text{ with } \mathbb{D}^e_{ijkl} = (K_u - \frac{2}{3}G)\delta_{ij}\delta_{kl} + G(\delta_{ik}\delta_{jl} + \delta_{il}\delta_{jk})\right),$$

in which K_u and G are the incompressibility modulus (or bulk modulus) for undrained conditions and the shear modulus, respectively. These two material properties are assumed to be constant over the whole column. The isotropic elasticity, fourth-order tensor $\mathbb{D}^e$ in (4a) is expressed in terms of these two material constants as well as of $\underline{\underline{\delta}}$ and $\mathbb{I}_S$, which are the second-order identity tensor and the symmetric fourth-order identity tensors, respectively. These identity tensors are also defined with the indicial notation in (4b) where δ_{ij}, the component of the second-order identity tensor, is the Kronecker symbol (equal to one if $i = j$ and equal to zero otherwise). The definition of the doubly-doted product ":" in (4a) is made clear in (4b) and thus $\underline{\underline{A}} : \underline{\underline{B}} = A_{ij} B_{ji}$ for any two second-order tensors $\underline{\underline{A}}$ and $\underline{\underline{B}}$. This general stress-strain relation for isotropic and elastic materials in (4) will be useful in the following sections and can be replaced by

$$\underline{\underline{\sigma}} = (K_u - \frac{2}{3}G)\underline{\underline{\delta}}\, \mathrm{tr}(\underline{\underline{\hat{\epsilon}}}) + 2G\underline{\underline{\hat{\epsilon}}} \,, \tag{5}$$

$$\left(\sigma_{ij} = (K_u - \frac{2}{3}G)\delta_{ij}\hat{\epsilon}_{kk} + 2G\epsilon_{ij}\right),$$

which is the usual presentation of the classical Hooke's law. Note in (5) the introduction of the trace operator with $\mathrm{tr}(\underline{\underline{\hat{\epsilon}}}) = \hat{\epsilon}_{kk}$.

The stress tensor obtained from (4) or (5), for the strain field in (3), is explicitly

$$\underline{\underline{\sigma}} = \left((K_u + \frac{4}{3}G)\underline{e}_z \otimes \underline{e}_z + (K_u - \frac{2}{3}G)(\underline{e}_x \otimes \underline{e}_x + \underline{e}_y \otimes \underline{e}_y)\right)\hat{u}_{,z} \,. \tag{6}$$

The application of the displacement method presented in Figure 2 has brought us to the step where a stress field has been created from the symmetric part of the gradient of the proposed K.A. displacement field. The next step is to question if this stress field is statically admissible (S.A.). A S.A. stress field satisfies the mechanical equilibrium conditions

$$\begin{aligned} \underline{\mathrm{div}}(\underline{\underline{\sigma}}) + \rho_s \underline{g} &= \underline{0} \quad \forall \underline{x} \in \Omega \, , \\ \Big(\sigma_{ij,j} + \rho_s g_i &= 0 \Big) \, , \end{aligned} \tag{7}$$

in which ρ_s and $\underline{g}$ are the material density and the gravity vector $-g\underline{e}_z$. Only the third of the set of three equations in (7) is non-trivial and, once the stress has been replaced by the unknown displacement function (6), it becomes

$$(K_u + \frac{4}{3}G)\hat{u}_{,zz} - \rho_s g = 0 \quad \forall z \in [0; H] \, . \tag{8}$$

This second-order differential equation requires two boundary conditions for the unknown function $u(z)$. The first condition is the zero displacement condition at the base. The second boundary condition is provided by the stress boundary conditions at the contact of the fluid layer ($z = H$). The stress vector is defined by $\underline{T} = \underline{\underline{\sigma}} \cdot \underline{n}$ for any surface oriented by its normal $\underline{n}$. At the boundary of the fluid layer, the normal is $\underline{e}_z$ and the stress vector obtained by multiplying (6) by $\underline{e}_z$ reads $(K_u + \frac{4}{3}G)u_{,z}\underline{e}_z$. For equilibrium, its normal component must be equal to the pressure at the depth e : $\underline{T}^d = -\rho_f g e \underline{e}_z$ in which ρ_f is the fluid material density. More generally, the S.A. stress field satisfies the boundary conditions

$$T_i = T_i^d \quad \forall \underline{x} \in \partial\Omega_{Ti} \, , \tag{9}$$

on the part of the boundary $\partial\Omega_{T_i}$ where some ($i = 1, 2$ or 3) of the components of stress vector has to be equal to the corresponding component of the applied distributed load. The dummy indices i or j in the definitions of the two parts of the boundary $\partial\Omega_{T_i}$ and $\partial\Omega_{u_j}$ are reminders that at a given point on the boundary one could specify the force density for certain directions (say i = 1,2) whereas the displacement is known in the remaining direction (j = 3).

The solution of the equilibrium equation (8) with the two boundary conditions given above is

$$\hat{u}(z) = \frac{\rho g}{K_u + \frac{4}{3}G}\Big(\frac{1}{2}z^2 - (\frac{\rho_f e}{\rho_s} + H)z\Big) \, . \tag{10}$$

To summarise the application of the displacements method: the K.A. displacement field which has been proposed lead us to a stress field which is piecewise continuous and thus differentiable. This stress field satisfies the equations of equilibrium over the domain Ω including the boundary conditions on the part of the boundary $\partial\Omega_{T_i}$ where some of the stress vector components must be equal to the applied surface force densities T_i^d (fluid pressure at top of column in this example). The set of all stress tensors having the same properties is the set of statically admissible stress fields[1] (S.A.) which is generally denoted

$$\mathcal{S}(\underline{g}, \partial\Omega_{T_i}, T_i^d) = \Big\{ \underline{\underline{\sigma}}^* | \text{ piecewise continuous and differentiable,} \qquad \text{satisfying equations (7) and (9)} \Big\} . \tag{11}$$

There is however a unique K.A. displacement field which is associated to a S.A. stress field by the poro-elasticity constitutive relations and it is the exact solution to the problem. It is thus concluded that equation (10) is the exact and unique solution, in terms of displacement, to the boundary value problem. The pore-fluid pressure has not been discussed so far. The material is assumed impermeable and the fluid mass is constrained in every pore so that the fluid pressure is determined locally. The pressure changes are linearly related to the trace of the stress tensor which we have determined here. Chapter 1 by F.K. Lehner provides the appropriate relation to compute them.

2.2 Variational Approach, Minimum of the Potential Energy

Variational methods assess the merits of S.A. or K.A. fields in providing good approximations to the exact solution in terms of energy. The central idea is to minimize the energy of the family of considered solutions knowing that the minimum is reached only by the exact solution. In this section, the minimum in potential energy for K.A. fields is presented, which is the basis of the introduction of the finite-element method in the next subsection.

As a preliminary to this discussion, a weak formulation of equilibrium is derived, referred to as the theorem of virtual work. It is constructed starting

[1] An additional equilibrium condition is often included to define the set of S.A. stress fields. It consists of the equilibrium condition across any potential stress discontinuity surface Σ_σ oriented by the normal $\underline{n}$ and reads $[\![\underline{\underline{\sigma}}]\!] \cdot \underline{n} = \underline{0}$. The double brackets denotes the jump in the argument defined as the difference $(\underline{\underline{\sigma}}^+ - \underline{\underline{\sigma}}^-)$ between the value of the argument on the two sides of the discontinuity, the $+$ side being pointed by the normal.

from any S.A. stress field denoted $\underline{\underline{\sigma}}^*$, and by taking the inner product of the equilibrium equation (7) by any K.A. field $\underline{\hat{u}}$ and integrating the resulting scalar over the domain

$$\int_\Omega (\underline{\mathrm{div}}(\underline{\underline{\sigma}}^*) + \rho_s \underline{g}) \cdot \underline{\hat{u}} \, dV = 0 \,, \tag{12}$$

$$\left(\int_\Omega (\sigma^*_{ij,j} + \rho_s g_i) \hat{u}_i \, dV = 0 \right) .$$

Integration by parts of the term $\underline{\mathrm{div}}(\underline{\underline{\sigma}}^*) \cdot \underline{\hat{u}}$ and application of the divergence theorem provides

$$\int_\Omega \underline{\underline{\sigma}}^* : \underline{\underline{\epsilon}}(\underline{\hat{u}}) \, dV = \int_\Omega \rho_s \underline{g} \cdot \underline{\hat{u}} \, dV + \int_{\partial\Omega} \underline{n} \cdot \underline{\underline{\sigma}}^* \cdot \underline{\hat{u}} \, dS \,, \tag{13}$$

having replaced the gradient of the displacement by the strain using the symmetry of the stress tensor. The work is virtual in the sense that any field, consistent with the boundary conditions, can be considered in (13) and not just the exact displacement field.

The first step in the construction of the minimum potential in terms of displacement is to propose the total potential energy of the column which is the difference between the internal energy W and the external energy Φ. The internal energy is the sum over the domain of the free energy or strain energy which, in the absence of initial stress and for linear elasticity, reads

$$w(\underline{\underline{\hat{\epsilon}}}) = \frac{1}{2}(K_u - \frac{2}{3}G) \ \mathrm{tr}(\underline{\underline{\hat{\epsilon}}})^2 + G \ \mathrm{tr}(\underline{\underline{\hat{\epsilon}^2}}) \,, \tag{14}$$

for any strain tensor $\underline{\underline{\hat{\epsilon}}}$ defined as $\underline{\underline{\epsilon}}(\underline{\hat{u}})$. This potential is differentiable and the partial derivative of the strain energy with respect to the strain $\partial w / \partial \underline{\underline{\epsilon}}$ is the stress defined in (5) The strain energy is a convex potential[2] meaning that

$$w\Big(\alpha \underline{\underline{\hat{\epsilon}_1}} + (1-\alpha)\underline{\underline{\hat{\epsilon}_2}}\Big) \leq \alpha w(\underline{\underline{\hat{\epsilon}_1}}) + (1-\alpha) w(\underline{\underline{\hat{\epsilon}_2}}) \quad \forall \alpha \in [0;1] \,, \tag{15}$$

for any pair of strains $\underline{\underline{\hat{\epsilon}_1}}$ and $\underline{\underline{\hat{\epsilon}_2}}$, as illustrated in Figure 3a. Convexity and differentiability of the strain energy leads to the property

$$w(\underline{\underline{\hat{\epsilon}_2}}) - w(\underline{\underline{\hat{\epsilon}_1}}) \geq \frac{\partial w}{\partial \underline{\underline{\epsilon}}}(\underline{\underline{\hat{\epsilon}_1}}) : (\underline{\underline{\hat{\epsilon}_2}} - \underline{\underline{\hat{\epsilon}_1}}) \,, \tag{16}$$

[2] Actually, a sufficient condition for convexity is that the modulus of elasticity E is positive and Poisson's ratio ν contained within the interval $]-1;1/2[$, two conditions always met in practice. Recall that $2E = (K_u - 2G/3)/(K_u + 4/3G)$ and $\nu = 3K_uG/(K_u + 4G/3)$.

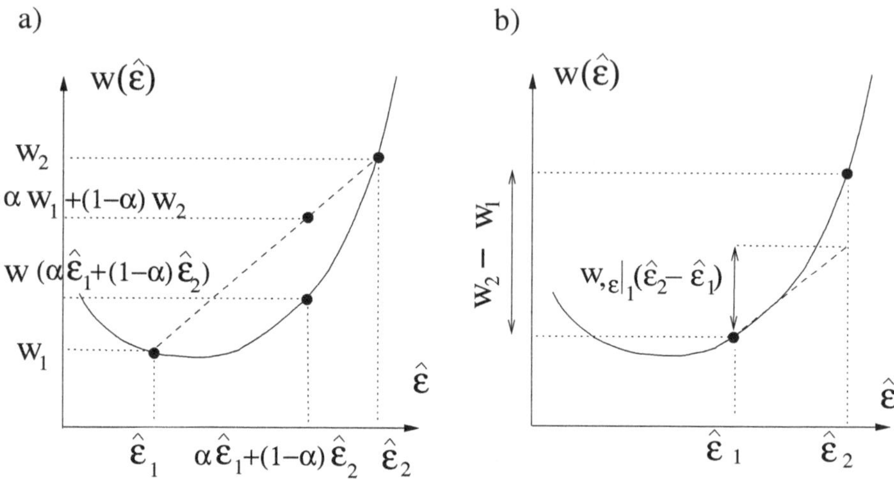

Figure 3. The convexity of the potential $w(\epsilon)$ in a 1D setting should help to understand the general definition of convexity in the six-dimensional strain space. The definition of the convexity (15) and the property (16) are illustrated in a) and b).

which is best illustrated in a 1D setting, Figure 3b. The internal energy for the boundary value problem is the sum of (14) over the domain

$$W(\underline{\hat{u}}) \equiv \int_\Omega w(\underline{\underline{\hat{\epsilon}}})\, dV = \int_\Omega \frac{1}{2}(K_u - \frac{2}{3}G)\ \mathrm{tr}(\underline{\underline{\epsilon}}(\underline{\hat{u}}))^2 + G\ \mathrm{tr}(\underline{\underline{\epsilon}}(\underline{\hat{u}})^2)\, dV\,, \quad (17)$$

for any K.A. displacement field $\underline{\hat{u}}$. The external work done by the K.A. displacement field over the external forces, characterised by the gravity vector $\underline{g}$ and the applied force density over the boundary, is

$$\Phi(\underline{\hat{u}}) = \int_\Omega \rho_s \underline{g} \cdot \underline{\hat{u}} dV + \sum_i \int_{\partial\Omega_T} T_i^d \hat{u}_i dS\,. \quad (18)$$

The sum in the right-hand side extends to the number of directions where the force is specified. In our example, the fluid pressure acting at the top of the column is the only applied force and

$$\Phi(\underline{\hat{u}}) = \int_0^h -\rho_s g \underline{e}_z \cdot \underline{\hat{u}} dz + \underline{T}^d \cdot \underline{\hat{u}}(h)\,. \quad (19)$$

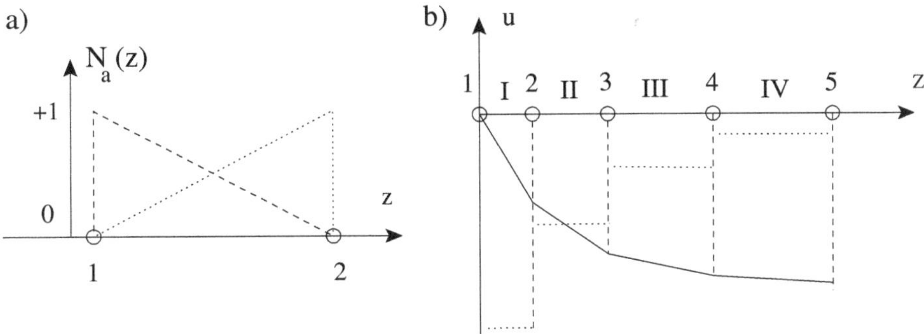

Figure 4. The two-noded element and its shape functions in a) and a typical piecewise linear K.A. displacement field in b) for the first example (poro-elastic column). The K.A. field has a constant gradient over each element (dotted segments) which is thus discontinuous over the column.

__Theorem__: The minimum in potential energy among all K.A. displacement fields is reached by the exact solution:

$$(W - \Phi)(\underline{u}) \leq (W - \Phi)(\underline{u}'), \quad \forall \underline{u}' \in \mathcal{C}(\partial\Omega_{u_i}, u_i^d). \tag{20}$$

The proof of this theorem is as follows. Apply the theorem of virtual work (13) to the exact solution $(\underline{u}, \underline{\underline{\sigma}})$ and then to the couple $(\underline{\hat{u}}, \underline{\underline{\sigma}})$ for any K.A. field $\underline{\hat{u}}$ and take the difference between the two resulting equalities to obtain

$$\int_\Omega \underline{\underline{\sigma}} : \left[\underline{\underline{\epsilon}}(\underline{u}) - \underline{\underline{\epsilon}}(\underline{\hat{u}})\right] dV = \int_\Omega \rho_s \underline{g} \cdot (\underline{u} - \underline{\hat{u}})\, dV + \int_{\partial\Omega} \underline{n} \cdot \underline{\underline{\sigma}} \cdot (\underline{u} - \underline{\hat{u}}) dS\,. \tag{21}$$

Since the two displacement fields $\underline{\hat{u}}$ and $\underline{u}$ are K.A., their difference is zero over $\partial\Omega_{u_i}$ and the surface integral in the right-hand side of (21) is non trivial only over $\partial\Omega_{T_i}$. This right-hand side is thus nothing but $\Phi(\underline{u}) - \Phi(\underline{\hat{u}})$ as can be seen from the definition in (18). Furthermore, in the left-hand side of (21), one recognizes in $\underline{\underline{\sigma}}$ the partial derivative of the strain energy with respect to the strain. The application of (16) means that this right-hand side is less then $W(\underline{u}) - W(\underline{u}')$. The combination of these results in (21) provides the inequality (20).

2.3 Finite-Element Approximation

The finite-element method is a systematic procedure to construct K.A. fields which approximate the exact solution with the minimum error in

Element	Local node 1	Local node 2
I	1	2
II	2	3
III	3	4
IV	4	5

Table 1. The connectivity table for the finite-element mesh presented in Figure 1b.

terms of energy. The first step of the procedure to construct a FE solution consists in discretising the domain by a set of elements defining the finite-element mesh, as illustrated in Figure 1b where four elements are numbered in Roman (total number of element, `numel` = 4). Each element covers the sub-domain Ω^e and is starting and ending with a node. The set of nodes are numbered globally from 1 to 5 in Figure 1b (total number of nodal points, `numnp` = 5). The construction of the K.A. displacement field is done element per element with the same systematic manner. For that purpose, the first and the last node of each element are labelled locally 1 and 2, respectively, Figure 4a. The nodes have thus two numberings, the first local (element level) to systematize the construction and the second numbering is global (structure scale). The link between the two numberings is found in the connectivity in Table 1.

The K.A. displacement field over the element e presented in Figure 4a varies linearly between the displacement $\hat{u}_1$ at node 1, of coordinate z_1 and the displacement $\hat{u}_2$ at node 2 of position z_2. The displacement at any point of coordinate z within the interval $[z_1, z_2]$ is thus

$$\hat{u}_h(z) = \hat{u}_1 \frac{z_2 - z}{l^e} + \hat{u}_2 \frac{z - z_1}{l^e} \quad , \quad l^e = z_2 - z_1 \, , \tag{22}$$

in terms of l^e, the element length. The subscript h marks the introduction of the spatial interpolation in terms of the characteristic size of the elements of the mesh, often denoted by the letter h. The displacement field could be also written locally as

$$\hat{u}_h(z) = \sum_{a=1}^{2} N_a(z) \hat{u}_a \, . \tag{23}$$

with the introduction of two functions $N_a(z)$, called shape functions and defined in (22). They are illustrated in Figure 4a by the dashed and the dotted segments. Each function $N_a(z)$ is zero outside the element, takes the value one at node a and zero at the other node. Note that the sum of the two shape functions at any point z is equal to one. These are general properties met by shape functions also in a 2D or a 3D spatial interpolation. Note as well that (23) may also be expressed in a matrix notation

$$\hat{u}_h(z) = [N(z)]_{1\text{x}2}\{\hat{u}\} \tag{24}$$

$$\text{with} \quad [N(z)] = [N_1(z), N_2(z)] \quad \text{and } \{\hat{u}\} = \left\{ \begin{array}{c} \hat{u}_1 \\ \hat{u}_2 \end{array} \right\},$$

in which the vector $\{\hat{u}\}$ contains the unknown, local nodal displacement. Brackets are used to define matrices with the number of lines and columns mentioned in subscript for sake of clarity whenever necessary. Column vectors are noted with curly brackets.

The construction of the local displacement interpolation over each element just presented leads to a K.A. displacement field over the whole structure which is continuous and piecewise linear between two consecutive nodes, as illustrated in Figure 4b. In our example, The displacement is thus expressed globally in terms of five unknown scalars which are the nodal displacements. Setting the displacement of node 1 to zero to satisfy the displacement boundary condition, any displacement field generated by our finite-element approximation becomes kinematically admissible and is an element of the set $\mathcal{C}_h(\partial\Omega_{u_i}, u_i^d)$. This set of generated admissible displacement fields is thus a subset of $\mathcal{C}(\partial\Omega_{u_i}, u_i^d)$. It is clear from the previous section, however, that the exact solution does not follow the piecewise interpolation proposed and is thus not within the subspace of search $\mathcal{C}_h(\partial\Omega_{u_i}, u_i^d)$. It is shown next how the finite-element solution will approximate the exact solution in order to minimize the error in terms of energy.

The objective is thus to find the element of $\mathcal{C}_h(\partial\Omega_{u_i}, u_i^d)$ which minimizes the total potential energy according to (20). This energy to be determined requires the computation of the strain $\hat{\epsilon}_h$, defined in (2), which may be written for each element as

$$\begin{aligned} \hat{\epsilon}_h(z) &= \sum_{a=1}^{2} N_{a,z}(z)\hat{u}_a\,, \qquad\qquad (25) \\ &= [B(z)]_{1\text{x}2}\{\hat{u}\} \quad \text{with} \quad [B(z)] \equiv [N_{1,z}(z), N_{2,z}(z)] = \frac{1}{l_e}[-1,+1]\,, \end{aligned}$$

introducing the matrix $[B]$. The strain approximation is the relative variation of the element length (by definition the nominal strain), $(u_2 - u_1)/l^e$, constant over each sub-domain. It is, as it is usually the case, discontinuous across the element boundaries as illustrated in Figure 4b with dotted segments. These discontinuities do not prevent the total energy of the mesh to remain bounded. This energy at the element level requires the knowledge of the stress. In our particular example, the vertical stress is associated to the approximated strain $\hat{\epsilon}_h(z)$ by $\hat{\sigma}_h = E_{1D}\hat{\epsilon}_h$ in terms of the 1D compaction elasticity modulus ($E_{1D} = K_u + 4G/3$). Consequently, the strain energy for a given element e is

$$W^e(\hat{u}_h) = \tfrac{1}{2}{}^t\{\hat{u}\}[k]^e_{2\text{x}2}\{\hat{u}\} \tag{26}$$

$$\text{with} \quad [k]^e \equiv \int_{\Omega^e} {}^t[B]E_{1D}[B]dV = \frac{E_{1D}}{l^e}\begin{bmatrix} 1 & -1 \\ -1 & 1 \end{bmatrix},$$

in which the t to the left of an array designates its transpose. The element strain energy in (26) is expressed in terms of the symmetric stiffness array of the element, the **2x2** array $[k]^e$. This local stiffness matrix has a zero determinant. The eigenmode $[1,1]\sqrt{2}/2$ associated to the zero eigenvalue corresponds to an arbitrary rigid translation, with consequences which will be seen by the end of this section. The total potential energy is the sum over the total number of elements `numel` within the mesh of the strain energy defined in (26)

$$W(\hat{u}_h) = \frac{1}{2}\sum_{e=1}^{\texttt{numel}} {}^t\{\hat{u}\}^e[k]^e\{\hat{u}\}^e \,. \tag{27}$$

The structure of (27) illustrates the contribution of each element to the total potential energy. A second notation is now proposed to provide a global view on this energy in terms of the nodal unknowns grouped in the column vector $\{\hat{U}\}$ which transpose reads

$${}^t\{\hat{U}\} \equiv (\hat{U}_1,\hat{U}_2,\hat{U}_3,\hat{U}_4,\hat{U}_5) = (\hat{u}^1_1,\hat{u}^1_2,\hat{u}^2_2,\hat{u}^3_2,\hat{u}^4_2) = (\hat{u}^1_1,\hat{u}^2_1,\hat{u}^3_1,\hat{u}^4_1,\hat{u}^4_2) \tag{28}$$

in which an uppercase letter is used for the global displacement and the superscript in $\hat{u}^e_a$ denotes the element number. Continuity of the displacement requires the condition $\hat{u}^e_2 = \hat{u}^{e+i}_1$ for the element $e = 1$ to `numel` $- 1$. The energy then reads

$$W(\hat{u}_h) = \frac{1}{2}{}^t\{\hat{U}\}[K]\{\hat{U}\}, \tag{29}$$

in which $[K]$, a `numnp` x `numnp` matrix, is the global, symmetric stiffness matrix

$$[K] = \begin{bmatrix} k^1_{11} & k^1_{12} & 0 & 0 & 0 \\ k^1_{21} & k^1_{22}+k^2_{11} & k^2_{12} & 0 & 0 \\ 0 & k^2_{21} & k^2_{22}+k^3_{11} & k^3_{12} & 0 \\ 0 & 0 & k^3_{21} & k^3_{22}+k^4_{11} & k^4_{12} \\ 0 & 0 & 0 & k^4_{21} & k^4_{22} \end{bmatrix}, \tag{30}$$

expressed in terms of the local stiffness components of the five elements (again, the superscript designates the element number). This construction of the global array from element contributions is called the assembly and is done efficiently by computers.

The second contribution to the total potential energy is due to the work done by the K.A. displacement field over the external forces. The work due to gravity over the four elements and the work done by the fluid layer of thickness e on top of the column adds to

$$\begin{aligned} \Phi(\hat{u}) &= \sum_{e=1}^{\texttt{numel}} {}^t\{\hat{u}\}\{f\}^e - \rho_f g e \hat{U}_5 \qquad (31) \\ \text{with} \quad \{f\}^e &\equiv \int_{\Omega^e} -\rho_s g\, {}^t[N] dV = -\frac{\rho_s g}{l_e} \left\{ \begin{array}{c} 1 \\ 1 \end{array} \right\}, \end{aligned}$$

in terms of the local work vector $\{f\}^e$. After assembly of the element contributions, the work done by the external forces is

$$\begin{aligned} \Phi(\hat{U}) &= {}^t\{\hat{U}\}\{F\} \qquad (32) \\ \text{with} \quad {}^t\{F\} &= [f^1_1, f^1_2 + f^2_1, f^2_2 + f^3_1, f^3_2 + f^4_1, f^4_2 - \rho_f g e]\,. \end{aligned}$$

The total potential energy becomes in terms of the global displacement unknowns

$$(W - \Phi)(\hat{U}) = \frac{1}{2} {}^t\{\hat{U}\}^t[K]\{\hat{U}\} - {}^t\{\hat{U}\}\{F\}, \tag{33}$$

and is now interpreted, in our particular problem, as a function of the 5 scalars which are the nodal displacements. The minimum within $\mathcal{C}_h(\partial\Omega_{u_i}, u_i^d)$ is attained as the partial derivative of $(W - \Phi)(\hat{u}_h)$ vanishes for all free variables $\hat{U}_a$. In this example, since the first displacement $\hat{U}_1$ is constrained to be zero because of the boundary condition, the minimization has to be

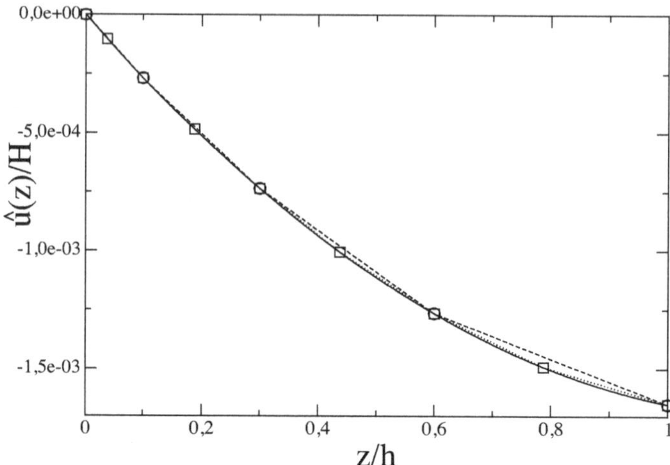

Figure 5. The displacement field for four (dashed curve) and eight (dotted curve) elements compared to the exact solution (solid curve) for the example of the 1D compaction of the poro-elastic column.

performed with respect to four displacement unknowns. The global system of linear equations to be solved is thus

$$[\tilde{K}]\{\tilde{U}\} = \{\tilde{F}\}. \tag{34}$$

in which the modified stiffness array $\tilde{K}$ corresponds to the one proposed in (30) amended of the first line and first column and the vector $\{\tilde{F}\}$ is the one defined in (32) omitting the first component. The vector $\{\tilde{U}\}$ thus does not contain the first nodal displacement which is known to be zero in our example. A few remarks on this system of equations are necessary. It is symmetric because the starting point of the FE solution construction is the minimization of the potential energy. The matrix is not full but banded because the finite-element interpolation is local and not global: the shape functions are only defined over each element and thus a nodal unknown influences only a sub-domain of Ω (two elements in our example). The global stiffness matrix in (30) obtained by assembly shares the zero determinant properties inherited from the local stiffness matrices with a global eigenmode which is also a rigid translation. The account of the displacement boundary conditions leading to (34) eliminates this mode from the solution.

The numerical application of this theory is now presented. The finite-

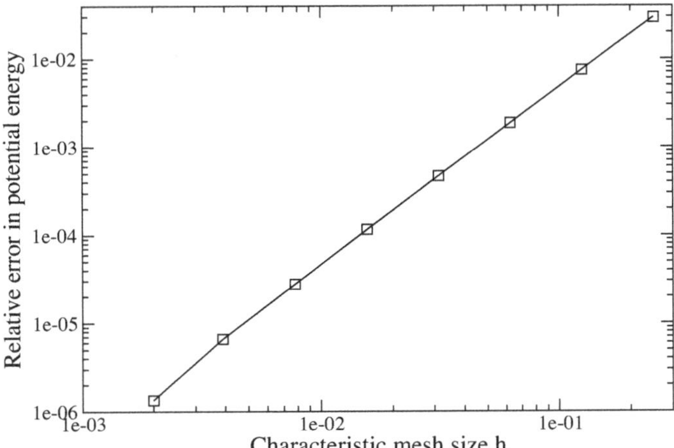

Figure 6. The relative error in potential energy as a function of the characteristic size h of each element for the first example of the compaction of the poro-elastic column (log-log plot).

element solutions are obtained for the following geometrical and material properties. The fluid of mass density $\rho_f = 1000\ kg/m^3$ occupies a layer of thickness $e = 50\ m$ while the sediments of mass density $\rho_s = 2300\ kg/m^3$ form a layer of thickness $H = 100\ m$. The sediments have an elastic modulus $E_{1D} = 1\ GPa$ and the gravity acceleration g is set to $10\ m/s^2$. The finite-element results are presented in Figures 5 and 6. The displacement for four and eight elements are presented in Figure 5 using the linear interpolation between the nodal value marked by circle and square, respectively. The approximate solutions are compared to the exact solution shown as a solid line. The accuracy of the approximate solutions is striking to the degree that the solutions for finer meshes are not presented here because they are difficult to differentiate in terms of displacement. This is partly due to an exceptional property of this 1D approximation which is that the exact solution is attained at each node. This property, called super-convergence and discussed by Strang and Fix (1973), does not extend to 2D nor 3D and is not explored further here. To assess the convergence of the finite-element approximations, the relative error in total potential energy has been plotted in Figure 6 as a function of the characteristic mesh size h (defined here as one over the number of elements). The rate of convergence is the slope of the line which approximates best this function in a log-log plot.

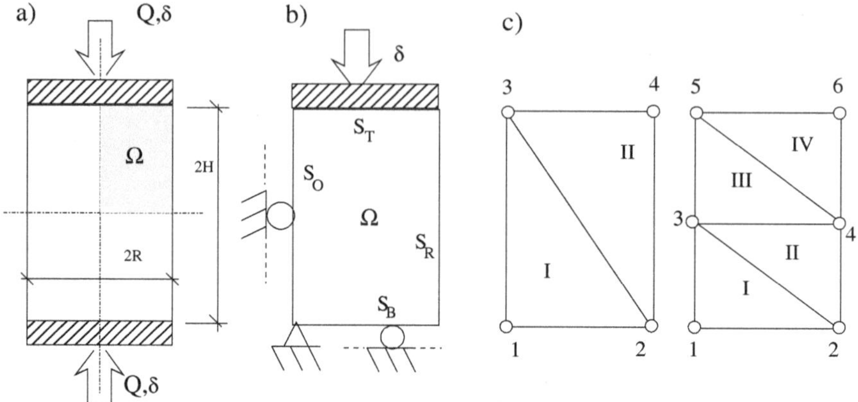

Figure 7. The geometry of the specimen with top and bottom platters constraining the compression in a). The region discretised for the finite-element approximation appears in grey. In b), the boundary conditions over the discretised domain along the tow symmetry axes and in c), two simple meshes composed of three-noded triangular elements.

The slope is found to be approximately +2. and could be compared to theoretical predictions (Ciarlet, 1978). Note that the convergence analysis was stopped at 500 elements since for finer meshes, the error is less then 10^{-6} corresponding to six significant digits in the value of the potential. The error cannot be measured for more than six significant digits since the maximum accuracy provided by a double precision computation, on most computers, is of seven significant digits.

3 The Second Example: Triangular Elements for 2D Linear Problems

The second example has for objective to introduce the simplest 2D element which is the three-noded triangle. The model problem is the constrained compression of a rectangular-shaped specimen of height $2H$ and width $2R$, Figure 7a. The structure is of infinite extent in the out-of-plane direction and the displacement field is only in-plane, corresponding to the hypotheses of plane-strain. The top and bottom surfaces of the specimen are clamped on two rigid platters which are displaced by δ so that the internal compressive deformation is not uniform. The symmetries of the geometry and of the loading allow to select the grey area in Figure 7a for defining the domain Ω for the finite-element analysis. The boundary conditions are shown in Figure 7b: the displacement along the top surface S_T is $-\delta \underline{e}_y$, the horizontal displacement and the tangential force density are zero along S_0 because of symmetry and the vertical displacement and the tangential force are zero along S_B for the same raison; finally, the applied force distribution (unit: Pa) along S_R is zero, conditions often described by comment "the surface is stress-free" in the French mechanics literature. Table 2 summarised these boundary conditions with the definition of the surfaces $\partial\Omega_{U_i}$ and $\partial\Omega_{T_i}$, introduced in the previous section.

Surface	Normal	Prescribed displ. ($\partial\Omega_{U_i}$)	Prescribed force density ($\partial\Omega_{T_i}$)
S_O	$-\underline{e}_x$	$u_x = 0$	$T_y = 0$
S_B	$-\underline{e}_y$	$u_y = 0$	$T_x = 0$
S_T	$\underline{e}_y$	$u_y = -\delta, u_x = 0$	
S_R	$\underline{e}_x$		$T_x = T_y = 0$

Table 2. The boundary conditions for the second FE example shown in Figure 7.

The selection of the displacement or force control on the boundary is identified in the third and fourth column. The second column provides the external normal, an information used for computing the stress vector.

3.1 Construction of the Finite-Element Approximation

Two finite-element meshes are proposed in Figure 7c to construct approximate solutions. They are composed of two and four three-noded triangular elements, respectively. Each element is similar to the one presented in Figure 8a with a local numbering from 1 to 3. The connectivity tables which link local and global numberings for the two meshes are provided in Table 3. Note that there are different ways to construct these tables, the only requirement being that the local numbering from 1 to 3 follows a clockwise rotation for each element.

The construction of the finite-element approximation follows the same procedure presented for the first example, starting with the displacement interpolation over each element. The local displacement field is

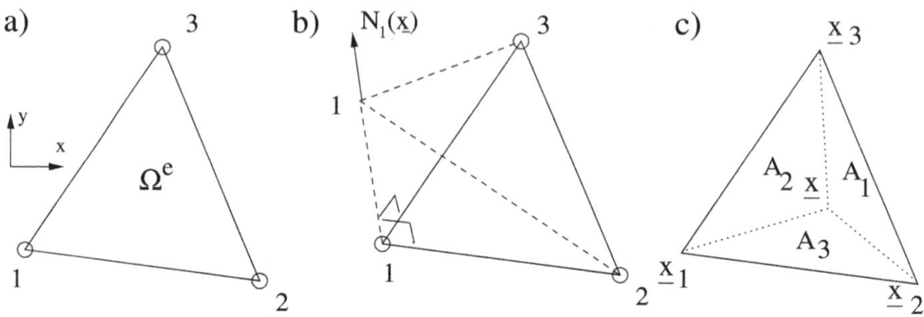

Figure 8. A three-noded triangular element covering the element domain Ω^e in a), the shape function associated to node 1 in b) and the definitions of the triangular surfaces to define the three shape functions, c).

Mesh	Element	Local node 1	Local node 2	Local node 3
1	I	1	2	3
	II	2	4	3
2	I	1	2	3
	II	2	4	3
	III	3	4	5
	IV	4	6	5

Table 3. The connectivity tables for the two finite-element meshes composed of triangular elements presented in Figure 7.

$$\underline{\hat{u}}_h(\underline{x}) = \sum_{a=1}^{3} N_a(\underline{x})\underline{\hat{u}}_a\,, \quad \{\hat{u}_h(\underline{x})\} = [N]_{2\times 6}\{\hat{u}\}\,, \tag{35}$$

$$\text{with} \quad [N] = \begin{bmatrix} N_1 & 0 & N_2 & 0 & N_3 & 0 \\ 0 & N_1 & 0 & N_2 & 0 & N_3 \end{bmatrix},$$

a linear combination of the nodal displacement denoted $\underline{\hat{u}}_a$ with the shape functions N_a. The second notation introduced in (35) generalises the matrix notation introduced in (23). The local nodal displacement vector $\{\hat{u}\}$ has now six components, which corresponds to the displacement in direction x and y (or 1 and 2) for node 1 to 3. To illustrate the structure of the shape functions N_a in (35), the first is presented in Figure 8b. This shape function is equal to one at node 1 and equal to zero at the two other nodes. The construction of these shape functions in terms of ratio of areas of the triangular regions defined in Figure 8c is such that $N_a(\underline{x}) = A_a(\underline{x})/A$ in which $A_a(\underline{x})$ is the area of the triangle with summits the point $\underline{x}$ and the two nodes which are not node a. The total area of the element is A. The area of each triangle is expressed in terms of the norm of the vector product of the vectors defining the sides of the triangles. With the Cartesian basis proposed in Figure 8a, the values of the four areas of interest read,

$$\begin{aligned} 2A_1 &= (x - x_3)(y_2 - y_3) - (y - y_3)(x_2 - x_3)\,, \\ 2A_2 &= (x_1 - x_3)(y - y_3) - (y_1 - y_3)(x - x_3)\,, \\ 2A_3 &= (x_2 - x_1)(y - y_1) - (y_2 - y_1)(x - x_1)\,, \\ 2A &= (x_2 - x_1)(y_3 - y_1) - (y_2 - y_1)(x_3 - x_1)\,. \end{aligned} \tag{36}$$

Having defined the interpolation of the displacement over a single element, we now discuss the construction of the global K.A. displacement field, taking the first of the two meshes in Figure 7c as an example. The transpose of the global nodal displacement vector $\{\hat{U}\}$ is composed of eight components and reads $(...\hat{U}_{a1}, \hat{U}_{a2}...)$ where the two components (direction 1 and 2 or x and y) of a generic node a are presented ($a = 1, ...\texttt{numel} = 2$). Any point within the domain Ω is part of the element I or II including their boundary and the displacement interpolation (35) applies. The boundary conditions in Table 2 constrain four components to be zero and two to be equal to $-\delta$ so that this global vector$\{\hat{U}\}$ reads

$$^t\{\hat{U}\} = (0, 0, \hat{U}_{21}, 0, 0, -\delta, 0, -\delta)\,. \tag{37}$$

There is a single unknown scalar, which is the third component of this global vector, the horizontal displacement of node 2. The space of kinematically admissible displacement $\mathcal{C}_h(\partial\Omega_{u_i}, u_i^d)$, which is then constructed for this mesh with two elements, is spanned by a single scalar $\hat{U}_{21}$.

The next step consists in defining the strain $\underline{\underline{\hat{\epsilon}}}_h(z)$ using the symmetric differential operator presented in (2). To this end, one needs the displacement gradient in (35) which relates the infinitesimal change in displacement

$d\underline{\hat{u}}_h(\underline{x})$ from a change in the material position $d\underline{x}$. This gradient is

$$\underline{\underline{\nabla u}}_h = \sum_{a=1}^{3} \underline{\hat{u}_a} \otimes \underline{\nabla N}_a \quad \text{with} \tag{38}$$

$$\begin{aligned}
\underline{\nabla N}_1 &= \frac{1}{2A}[(y_2 - y_3)\underline{e}_x + (x_3 - x_2)\underline{e}_y]\,, \\
\underline{\nabla N}_2 &= \frac{1}{2A}[(y_3 - y_1)\underline{e}_x + (x_1 - x_3)\underline{e}_y]\,, \\
\underline{\nabla N}_3 &= \frac{1}{2A}[(y_1 - y_2)\underline{e}_x + (x_2 - x_1)\underline{e}_y]\,.
\end{aligned}$$

The symmetric part of this displacement gradient is now used to construct the column vector $\{\hat{\epsilon}_h\}$ which contains four components $\hat{\epsilon}_{11}$, $\hat{\epsilon}_{22}$, $\hat{\epsilon}_{33}$ and $2\hat{\epsilon}_{12}$. The relation between the strain vector and the nodal displacement is classically presented with the B operator or matrix, generalizing the concept introduced in (25). For plane-strain problems[3], it is defined by

$$\{\hat{\epsilon}_h\} = [B]\{\hat{u}\}\,, \quad [B] \equiv \begin{bmatrix} N_{1,1} & 0 & N_{2,1} & 0 & N_{3,1} & 0 \\ 0 & N_{1,2} & 0 & N_{2,2} & 0 & N_{3,2} \\ 0 & 0 & 0 & 0 & 0 & 0 \\ N_{1,2} & N_{1,1} & N_{2,2} & N_{2,1} & N_{3,2} & N_{3,1} \end{bmatrix}, \tag{39}$$

in which the B operator for the triangular element of interest is

$$\frac{1}{2A}\begin{bmatrix} y_2 - y_3 & 0 & y_3 - y_1 & 0 & y_1 - y_3 & 0 \\ 0 & x_2 - x_3 & 0 & x_3 - x_1 & 0 & x_1 - x_3 \\ 0 & 0 & 0 & 0 & 0 & 0 \\ x_2 - x_3 & y_2 - y_3 & x_3 - x_1 & y_3 - y_1 & x_1 - x_3 & y_1 - y_3 \end{bmatrix}. \tag{40}$$

The stress vector contains the four components $\hat{\sigma}_{11}$, $\hat{\sigma}_{22}$, $\hat{\sigma}_{33}$ and $\hat{\sigma}_{12}$ and is related to the strain vector by

$$\begin{aligned}
\{\hat{\sigma}_h\} &= [D_u^e]_{4\times 4}\{\hat{\epsilon}_h\} \\
\text{with} \quad [D_u^e] &\equiv \begin{bmatrix} K_u + \frac{4}{3}G & K_u - \frac{2}{3}G & K_u - \frac{2}{3}G & 0 \\ K_u - \frac{2}{3}G & K_u + \frac{4}{3}G & K_u - \frac{2}{3}G & 0 \\ K_u - \frac{2}{3}G & K_u - \frac{2}{3}G & K_u + \frac{4}{3}G & 0 \\ 0 & 0 & 0 & G \end{bmatrix},
\end{aligned} \tag{41}$$

[3] The reason to keep the third component of the strain, which is trivial for plane-strain, is that the size of the B operator is then the same for plane-stress or axi-symmetric problems.

in which the material stiffness array $[D_u^e]$ is derived from Hooke's law in (4). This matrix relation generalizes to 2D the scalar relation between the single stress component, the single strain component and the modulus E_{1D} in the previous example. The strain energy for a single element is now easily computed

$$W^e(\underline{\hat{u}}) = \frac{1}{2} {}^t\{\hat{u}\}[k]^e_{6x6}\{\hat{u}\} \quad \text{with} \quad [k]^e = \int_{\Omega^e} {}^t[B][D_u^e][B]dV\,, \tag{42}$$

which is similar in structure to (26) and where the local stiffness matrix $[k]^e$ is now a 6x6 matrix. The expression for the external work originates from (18) and, restricted to a single element, with the introduction of the displacement interpolation, reads

$$\begin{aligned} \Phi(\hat{u})^e &= {}^t\{\hat{u}\}\{f\}^e \\ \text{with} \quad \{f\}^e &= \int_{\Omega^e} \rho_s {}^t[N]\{g\}dV + \int_{\partial\Omega^e} {}^t[N]\{T^d\}dV\,, \end{aligned} \tag{43}$$

in which the vectors $\rho_s\{g\}$ and $\{T^d\}$ are the external body force vector and the density of applied forces if $\partial\Omega^e$ is part of the boundary of the domain where tractions are prescribed. The external work is zero in this particular example because gravity is disregarded and there is no force density prescribed on the boundary (see Table 2).

The rest of the construction of the finite-element approximation is similar to the one presented for the first example. Assembly of the element contributions in strain energy and external work leads to expressions of same structure as (29) and (32). Minimization with respect to the K.A. displacement leads to the symmetric system of equations of type (34) which has to be solved. This minimization and the solution of the resulting system of equations is discussed in further details, again for the first of the two meshes of Figure 7c. If all displacements were unconstrained, then the global system would be composed of eight equations. In this particular example, minimization is only done for $\hat{U}_{21}$, the third component of the vector which corresponds to the horizontal displacement of node 2. The only equation to be solved is thus

$$K_{33}\hat{U}_{21} + K_{36}(-\delta) = 0 \quad \text{with} \quad K_{33} = k^1_{33} + k^2_{11}\,, \quad K_{36} = k^1_{36} + k^2_{13}\,, \tag{44}$$

in which the components of the global stiffness have been expressed in terms of the components of the two local stiffness matrices.

The construction of a FE approximation for more complex meshes follows the same rules as the ones presented above. Results are presented next for a regularly spaced mesh composed of 400 triangles.

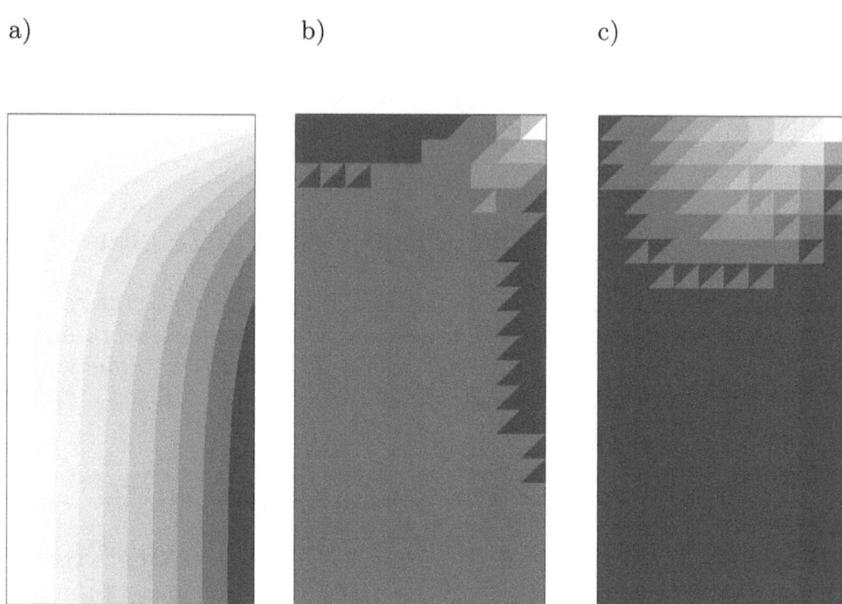

Figure 9. Results for the plane-strain, constrained compression of a rectangular block. Ten isocontours of the first component of the displacement in a), distribution of σ_{22} and σ_{12} in b) and c). The spectrum from white to black covers the distribution of each quantity over the range $[0; 1.28 \times 10^{-5}]$ m, $[-14.1; -9.9]$ MPa and $[-2.1; 0.09]$ MPa, respectively.

3.2 Analysis of the Results

The results are obtained for an elastic material having an undrained Young's modulus of 10 GPa and a Poisson's ratio of 0.2. The specimen width R and height H are 5 and 10 cm, respectively. The top displacement is $\delta = 0.1$ mm such that the nominal strain is 0.1%. The mesh is composed of 10 by 20 square cells composed each of two three-noded triangles with a common inclined boundary at $+45°$. Results are presented in Figure 9 in terms of the first component of the displacement field, and two components of the stress, σ_{22} and σ_{12}.The contouring is obtained with the graphic program GiD (2009).

The first displacement varies from zero on the vertical axis and along the top boundary to the maximum at the stress-free boundary close to the

1-axis. The constraint due to the top platter is clear and one needs to be half down the specimen to get an approximately constant gradient in the 1-direction.

For the two stress components, it has been preferred to stay with the hypothesis of the constant strain construction of the finite-elements and a constant value has been assigned to each element. The value of σ_{22} along the horizontal axis is close to the exact solution for unconstrained compression which is $-\delta/HE/(1-\nu^2)$. Note the strong gradient close to the upper right corner of the specimen which is due to the discontinuity in the tangential applied force distribution on the boundary. There is indeed an applied tangential force below the platter whereas the vertical surface is stress free. The consequence of such discontinuous stress loading conditions in terms of stress field discontinuity has been studied analytically (e.g. Timoshenko and Goodier, 1934). Finite-elements are not designed to capture the exact nature of this discontinuity but can provide a practical approximation with a sufficiently fine mesh. Note that the shear stress is also discontinuous at the top right corner. It is close to zero in the bulk of the specimen, away from the top boundary.

4 Quadrature Rules: Gauss and Lobatto

During the presentation of the two first examples of finite elements, the question came to compute a quadrature of the type

$$\int_{-1}^{+1} f(\xi) d\xi \,, \tag{45}$$

for either the stiffness array or the force vector. These computations are at the element level and we assume in this section a 1D setting over the interval $[-1;+1]$ on the ξ-axis. For the simpler elements and uniform material properties, the integrals of type (45) are computed analytically but the calculation become cumbersome or impossible for more complex elements or for non-linear material responses. It is then necessary to determine the best approximation possible of (45) by numerical means

$$\int_{-1}^{+1} f(\xi) d\xi \simeq \sum_{i=1}^{\mathtt{nquad}} f(\xi_i) w_i \,. \tag{46}$$

The function is sampled at pre-determined positions ξ_i, the **nquad** quadrature points, and the weighted sum by the coefficient w_i provides the approximation. Figure 10 illustrates the construction of this approximation for an arbitrary function defined on the interval $[-1,+1]$. The function f

is estimated a three points ξ_i which are centred on three sub-intervals of width, the weight w_i. The approximation of the quadrature is the sum of the three products between the function estimates and the corresponding weights. It corresponds to the sum of the areas which are shaded in Figure 10. The choice of the point positions and the weights is done to obtain the best possible approximation which thus depends on the quadrature rule which is selected. Two rules are considered here: Gauss and Lobatto. The first of the two is now examined.

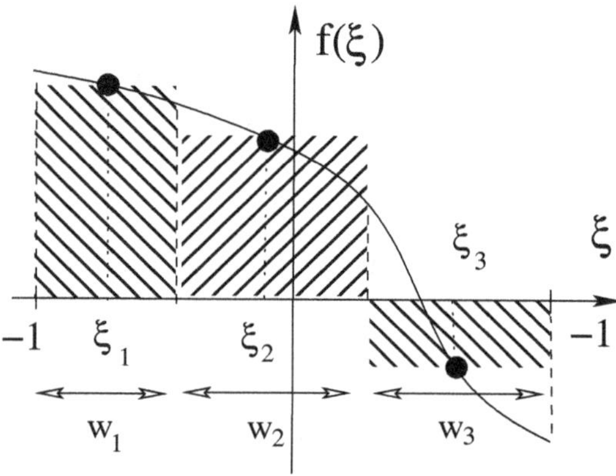

Figure 10. Illustration of the interpretation of the weighted sum (46) to approximate the surface below the arbitrary function f defined over the interval $[-1;+1]$.

The question is thus how to choose the weights and the positions of the quadrature points to obtain the best estimate of (45). Consider the case of **nquad** $=1$ for which there are two unknowns w_1 and ξ_1. They are determined by requiring that any linear polynomial $a_0 + a_1\xi$ should be integrated exactly. The error between the exact solution and the numerical approximation is

$$\mathcal{E} = \int_{-1}^{+1} a_0 + a_1\xi d\xi - w_1(a_0 + \xi_1 a_1)\,. \tag{47}$$

Setting this error to zero for any value of a_0 and a_1 leads to the two equations

$$\xi_1 = 0\,, \quad w_1 = 2\,. \tag{48}$$

The same procedure applies for arbitrary values of **nquad**. Any polynomial $a_i\xi^i$ of order **2nquad** -1 (i varies from 0 to **2nquad** -1) should be integrated exactly. The error $\mathcal{E}$ is then

$$\mathcal{E} = \sum_{i=0}^{\mathtt{2nquad}-1} \frac{a_i}{i+1}[1+(-1)^i] - \sum_{j=1}^{\mathtt{nquad}} w_j \sum_{i=0}^{\mathtt{2nquad}-1} a_i \xi_j^i\,, \tag{49}$$

after integration of each term of the polynomial. The error should be zero for any choice of the coefficient a_i, providing **2nquad** $-$ 1 equations. The first equation is due to the arbitrariness of the scalar a_0 and constrains the sum of all the weights to be equal to the length of the segment

$$\sum_{i=1}^{\mathtt{nquad}} w_i = 2\,. \tag{50}$$

The even-numbered equations are enforced trivially if the integration points are distributed symmetrically with respect to the origin of the ξ-axis and the weight are then identical at the same distance from this origin. The odd-numbered equations provides the values of ξ_i and w_i. Explicit values for the positions and the weigths of Gauss's quadrature rules are found in standard text books (e.g. Zienkiewicz, 1977). The important result to remember is that **nquad** quadrature points are necessary to integrate exactly a polynomial of order **2nquad - 1**.

The construction is more complex for **Lobatto's quadrature** rule and the interested reader is invited to read the book of Hildebrand (1974). Lobatto's rule differs from Gauss's rule by requesting that two quadrature points are found at $\xi = -1$ and $\xi = +1$. These two additional constraints have the consequence that **nquad** quadrature points can only integrate exactly a polynomial of order **2nquad - 3**.

The construction of these two quadrature rules was proposed for the segment $[-1;+1]$ on the ξ-axis. Keeping with this 1D setting, one could imagine that the finite-elements are placed on a curve with curvilinear coordinate s as shown in Figure 11. The elements are placed irregularly on that curve and we choose to concentrate on an arbitrary element which is bounded by the nodes numbered locally 1 and 2 of coordinate s_1 and s_2. There exists a mapping between the segment $[-1;+1]$ and the curved segment occupied by that element

$$\begin{aligned} \phi : [-1;+1] &\rightarrow [s_1;s_2] \\ \xi &\mapsto s = \phi(\xi)\,. \end{aligned} \tag{51}$$

This mapping is between the segment $[-1;+1]$ referred to as the computational domain and the physical domain $[s_1;s_2]$ occupied by the element. It

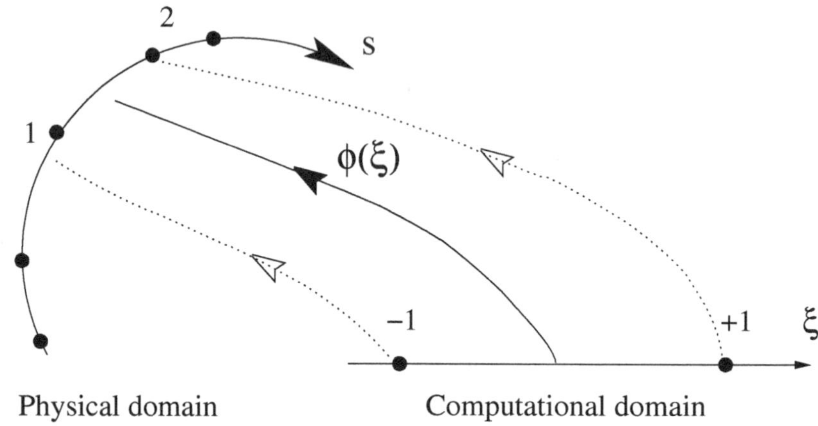

Figure 11. The mapping between the 1D computational domain and the physical domain for an arbitrary element bounded by two nodes numbered locally 1 and 2.

is assumed invertible for all ξ such that the Jacobian of the transformation $J = ds/d\xi = d\phi/d\xi$ is strictly positive. For the 1D linear element of the first example, this mapping is linear and the Jacobian is $l_e/2$ which is the ratio of the length of the element measured in the physical domain to the length of the segment over the ξ-axis. In general, any quadrature over the physical domain is then written equivalently over the computational domain

$$\int_{s_1}^{s_2} f(s)ds = \int_{-1}^{+1} f \circ \phi(\xi) J d\xi \,, \tag{52}$$

with the notation $f \circ g(\xi) = f(g(\xi))$ for the composition of two applications f and g. The integral in (52) is then approximated

$$\int_{s_1}^{s_2} f(s)ds \simeq \sum_{i=1}^{\mathtt{nquad}} f \circ \phi(\xi_i)\, w_i J(\xi_i) \,, \tag{53}$$

corresponding to the sum of the function f evaluated at the series of points of coordinate $\phi(\xi_i)$ in the physical space and weighted by the scalar w_i times the Jacobian of the transformation estimated at the equivalent position ξ_i in the computational domain. A quadrature over any arbitrary element in the physical space can thus systematically be mapped to the same compu-

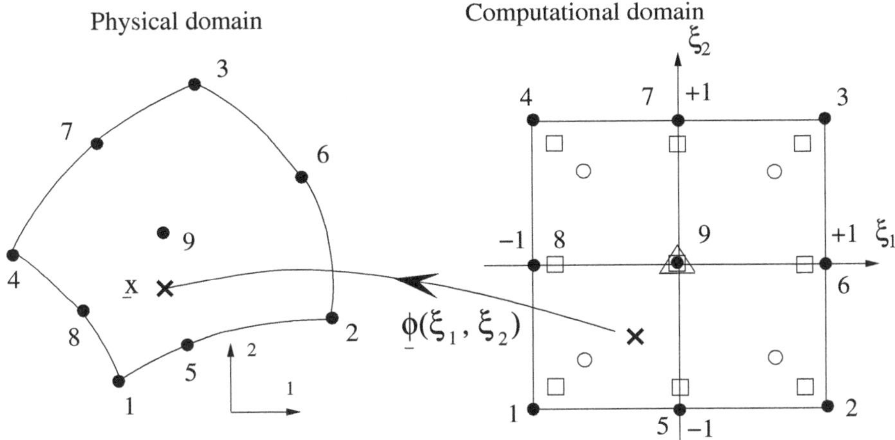

Figure 12. A four- to nine-noded Lagrange or serendipity element with its local nodal numbering and its mapping from the computational domain to the physical domain. The 1×1, 2×2 and 3×3 Gauss'quadrature points are presented by the open triangle, circles and squares, respectively.

tational domain defined by the segment $[-1;+1]$ over which the numerical quadrature rule providing the best approximation has been established. This systematic mapping is efficiently conducted with FE codes. We shall see next how this mapping and numerical quadrature are defined in a 2D setting for the elements of the Lagrange and the serendipity families.

5 Lagrange Elements and the Serendipity Family

One of the most popular family of 2D elements is the Lagrange family which contains elements having up to nine nodes, Figure 12. If the ninth, central node is omitted, the resulting set of elements having a maximum of eight nodes is called the serendipity family. The first objective of this section is to discuss the construction of the mapping between the physical and the computational domains as well as the displacement interpolation for these two families. The second objective is to analyse the convergence properties of 2D elements. Most of the properties which are discussed here are more general and apply to most finite elements.

5.1 The Spatial Interpolation

The mapping between the computational domain and the physical domain occupied by a single element is defined by

$$\underline{x}(\underline{\xi}) = \sum_{a=1}^{\mathtt{nel}} N_a(\underline{\xi})\underline{x}_a\,, \tag{54}$$

in terms of the shape functions N_a and the position $\underline{x}_a$ of the `nel` nodes (4 to 9) in the physical domain. Their position in the computational domain is defined in Figure 12 in a plane spanned by the ξ_1- and ξ_2-axes. The computational domain corresponds to the square $\Omega^c = [-1;+1] \times [-1;+1]$. The first four shape functions in (54) are

$$N_1(\xi_1,\xi_2) = \tfrac{1}{4}(1-\xi_1)(1-\xi_2)\,, \quad N_2(\xi_1,\xi_2) = \frac{1}{4}(1+\xi_1)(1-\xi_2)\,, \tag{55}$$
$$N_3(\xi_1,\xi_2) = \tfrac{1}{4}(1+\xi_1)(1+\xi_2)\,, \quad N_4(\xi_1,\xi_2) = \frac{1}{4}(1-\xi_1)(1+\xi_2)\,.$$

They satisfy each or collectively two basic rules:

(i) the shape function N_a takes the value of 1 at node a and of 0 at other nodes.

(ii) the sum of all shape functions is one at any point within the computational domain.

These two rules are at the basis of the construction of the five other shape functions and of the modification of (55) if more than four nodes are considered. For example, consider node 5 which is between nodes 1 and 2 at the coordinate ($\xi_1 = 0, \xi_2 = -1$). The shape function N_5 should contain the quadratic term $(1-\xi_1)(1+\xi_1)$ to satisfy rule *(i)*. Multiply this first term by $(1-\xi_2)$ to make sure that the shape function is zero at nodes 3 and 4. Normalize the product by 1/2 to make sure that N_5 is one at node 5. Once this new function is set, we need to correct the function N_1 to N_4 in (55) to make sure that they are zero at node 5. For example, $N_1(\xi_1 = 0, \xi_2 = -1) = 1/2$ in (55) so the correction to the first shape function should be $-1/2N_5$ so that the first shape function for the five-noded element becomes $N_1 - 1/2N_5$. The other shape function (node 2 to 4) should be amended similarly. The addition of any new node is done with the same strategy, first find the new shape function which is one at that node and zero at others and, second, modify the previously defined shape functions to account for the presence of this new node. Rule *(ii)* is then satisfied for any element having from 4 to 9 nodes. It is this reasoning which is at the base of the construction of Table 4.

	Basic	+ node 5	+ node 6	+ node 7	+ node 8	+ node 9
1	N_1	$-\frac{1}{2}N_5$			$-\frac{1}{2}N_8$	$-\frac{1}{4}N_9\times$ $(1-\delta_5-\delta_8)$
2	N_2	$-\frac{1}{2}N_5$	$-\frac{1}{2}N_6$			$-\frac{1}{4}N_9\times$ $(1-\delta_5-\delta_6)$
3	N_3		$-\frac{1}{2}N_6$	$-\frac{1}{2}N_7$		$-\frac{1}{4}N_9\times$ $(1-\delta_6-\delta_7)$
4	N_4			$-\frac{1}{2}N_7$	$-\frac{1}{2}N_8$	$-\frac{1}{4}N_9\times$ $(1-\delta_7-\delta_8)$
5		$N_5 = \frac{1}{2}\times$ $(1-\xi_1^2)\times$ $(1+\xi_2)$				$-\frac{1}{2}N_9$
6			$N_6 = \frac{1}{2}\times$ $(1+\xi_1)\times$ $(1-\xi_2^2)$			$-\frac{1}{2}N_9$
7				$N_7 = \frac{1}{2}\times$ $(1-\xi_1^2)\times$ $(1+\xi_2)$		$-\frac{1}{2}N_9$
8					$N_8 = \frac{1}{2}\times$ $(1-\xi_1)\times$ $(1-\xi_2^2)$	$-\frac{1}{2}N_9$
9						$N_9 = \frac{1}{2}\times$ $(1-\xi_1^2)\times$ $(1-\xi_2^2)$

Table 4. The final expression for the shape function of node 1 to 9 is defined on each line and is obtained by collecting the terms corresponding to the additional nodes presented in each column. The notation δ_a means 0 or 1 if node a is absent or present.

The Jacobian of the mapping between the computational and the physical domain is the determinant of the gradient of this transformation. The gradient relates the infinitesimal changes in position $d\underline{x}$ and $d\underline{\xi}$

$$d\underline{x}(\underline{\xi}) = \underline{\underline{\nabla_{\xi} x}} \cdot d\underline{\xi}\,, \qquad \text{with} \quad \underline{\underline{\nabla_{\xi} x}} = \sum_{a=1}^{\mathtt{nel}} \underline{x}_a \otimes \underline{\nabla_{\xi} N_a(\underline{\xi})}\,, \tag{56}$$

obtained by linearisation of (54) and with ∇_{ξ} being the gradient operator in the computational domain. The Jacobian is thus the determinant of the second-order tensor $\underline{\underline{\nabla_{\xi} x}}$ (components $\partial x_i/\partial \xi_j$) which is the sum over the nodes of the dyadic products found in the right-hand side of (56). It is defined in a matrix notation as

$$J(\underline{\xi}) = \det\left(\begin{bmatrix} \dfrac{\partial x_1}{\partial \xi_1} & \dfrac{\partial x_1}{\partial \xi_2} \\ \\ \dfrac{\partial x_2}{\partial \xi_1} & \dfrac{\partial x_2}{\partial \xi_2} \end{bmatrix}\right). \tag{57}$$

This Jacobian should be strictly positive at any point of the computational domain. It is used to transform any integral computed in the physical domain over an element of volume Ω^e back to the computational domain Ω^c

$$\int_{\Omega^e} f(\underline{x})dV = \int_{\Omega^c} f \circ \phi(\underline{\xi}) J(\underline{\xi}) dV_c\,. \tag{58}$$

Such integral will require a numerical quadrature which is best conducted with one of the quadrature rules described in the previous section in a 1D setting. The generalization of the numerical quadrature in 2D is as follows. Select first the 1D quadrature rule, say 1, 2 or 3 quadrature points. Second, adopt the same rule in the two dimensions of the computational space so that (58) becomes

$$\int_{\Omega^e} f(\underline{x})dV \simeq \sum_{i=1}^{\mathtt{nquad}} f \circ \phi(\underline{\xi}_i) J(\underline{\xi}_i) w_i\,, \tag{59}$$

in which the nquad quadrature points are 1×1, 2×2 or 3×3. Their position $\underline{\xi}_i$ are presented by the open triangle, circles and squares, respectively, in Figure 12. The position $\underline{\xi}_i$ are obtained by combining the various 1D coordinates and the weight w_i is the product of the 1D weights used in the two spatial directions. A quadrature of order n requires $n \times n$ quadrature points and provides an exact integration of a complete polynomial in ξ_1 and ξ_2 of order $2n-1$. Typically, for linear elasticity and elasto-plasticity problems, 2×2 quadrature points are used for 4 noded-elements and 3×3 for elements having a number of nodes between 5 and 9.

5.2 Displacement Interpolation and the Strain Operator B

The construction of the finite-element approximation of the elements of the Lagrange and serendipity families for linear elasticity is identical to the one presented for the triangular, three-noded elements in the previous section, except for the displacement interpolation and the strain operator which are now discussed.

It is first assumed that the spatial and the displacement interpolation are constructed with the same nodes and shape functions. In that instance the element is said to be *isoparametric*. The displacement interpolation over each element is thus

$$\underline{u}_h(\underline{x}) = \sum_{a=1}^{\mathtt{nel}} N_a\big(\phi^{-1}(\underline{x})\big)\underline{u}_a \,, \tag{60}$$

in which $\underline{u}_a$ is the nodal displacement and N_a the shape functions defined in (55) and Table 4. The matrix notation is identical to the one proposed in (35) with the $[N]$ array being now of dimension $2 \times 2\mathtt{nel}$. The strain operator or matrix $[B]$ defined in (40) is generalised to read

$$[B]_{4\times 2\mathtt{nel}} \equiv \left[\;\cdots\; \begin{pmatrix} N_{a,1} & 0 \\ 0 & N_{a,2} \\ 0 & 0 \\ N_{a,2} & N_{a,1} \end{pmatrix} \;\cdots\; \right] . \tag{61}$$

It is composed of nel blocks of size 4×2 which have the same structure for each node and a single block for an arbitrary node a is presented in (61). The gradient $N_{a,j}$ in (61) is with respect to the coordinate x_j in the physical domain. It is obtained by the chain rule of differentiation

$$\frac{\partial N_a}{\partial x_j} = \frac{\partial N_a}{\partial \xi_i}\frac{\partial \xi_i}{\partial x_j} \,, \tag{62}$$

and requires the inversion of the spatial gradient $\underline{\underline{\nabla_\xi x}}$ defined in (56) which is done numerically at every quadrature point.

5.3 Convergence Properties

The question is whether or not we are guaranteed to converge to the exact solution as the mesh size is tending towards zero. Section 2 helped us to understand the conditions in terms of energy in a 1D setting. More generally, convergence properties of the finite-element method for elliptic problems (linear elasticity and equilibrium is a typical example) have been given a sound mathematical basis (Ciarlet, 1978). Here, for 2D and 3D, we propose to recourse to three sufficient conditions which are discussed most often in the engineering literature and are appropriate for the two families of interest. The global displacement approximation is assumed continuous and piecewise continuously differentiable. The three conditions are

(i) the shape functions are C_1 (continuously differentiable) over each element interior,

(ii) the shape functions are continuous across the common boundary of two elements,

(iii) the shape functions are complete polynomials.

Elements which satisfy properties *(i)* and *(ii)* are said to be conforming or compatible. This is the case of the three-noded triangular elements found in section 4 and of the elements of the Lagrange and serenpidity families constructed in this section. Note that property (ii) ensures that the energy computed over the discretised domain is finite. Condition *(iii)* is proposed so that a global displacement field described by a polynomial function of the spatial coordinates can be reproduced at the local level with proper choice of the nodal variables. To understand this requirement, consider the limit of the characteristic mesh size h tending to 0. In that limit, the solution is more and more uniform at the local level leading to, approximately, a constant strain over the elements. Each element should thus have the basic property to reproduce a state of constant strain. The displacement field should be affine to represent this homogeneous strain with additional kinematics to represent a rigid, infinitesimal rotation and translation. We shall now examine this desired property of the displacement field for the Lagrange and serendipity families. Consider thus the exact displacement field described by

$$\underline{u}(\underline{x}) = \underline{\underline{A}} \cdot \underline{x} + \underline{b}\,, \tag{63}$$

in which $\underline{\underline{A}}$ combines the infinitesimal rotation with some constant stretching and the vector $\underline{b}$ is the translation. The displacement gradient is $\underline{\underline{\nabla u}} = \underline{\underline{A}}$ and the linearised strain the symmetric part of $\underline{\underline{A}}$. Consider now that the nodal displacements of any element are prescribed according to (63)

$$\underline{u}_a = \underline{\underline{A}} \cdot \underline{x}_a + \underline{b}\,, \tag{64}$$

in which $\underline{x}_a$ is the node position in the physical domain. The question is whether we can reproduce the same exact gradient at any point within this element. The displacement interpolation is given by (60) which combined with (64) provides

$$\underline{u}_h(\underline{x}) = \underline{\underline{A}} \cdot \sum_a^{\texttt{nel}} N_a \underline{x}_a + \underline{b} \sum_a^{\texttt{nel}} N_a \,. \tag{65}$$

Since the elements are isoparametric, the first sum in the right-hand side of (65) is the spatial position $\underline{x}_h$. The linearization of (65) $d\underline{u}_h = \underline{\underline{A}} \cdot d\underline{x}_h$ reveals that the exact gradient $\underline{\underline{A}}$ is well captured. However, for the displacement to be correct in (65), one requires because of the translation

$$\sum_a^{\texttt{nel}} N_a(\underline{x}) = 1 \,, \tag{66}$$

at every point within the element. This condition was proposed as condition (ii) after (55) and is respected for any element of the two families of interest here.

A final remark in this section concerns the elements which are incompatible. They could satisfy the completeness condition as a block of elements and not as single element. It has been proved that this collective property is sufficient to ensure convergence. This convergence property of a block or patch of elements is due to Irons (Bazeley et al., 1965). Incompatible elements verifying this condition are said to pass the patch test. It is this test which is explored in further details in the problem section.

6 Plasticity for Small Perturbations and Finite Transformation

The objective of this section is to introduce the finite-element method for problems with non-linearities which are either due to the material properties or to the geometry. The material non-linearities which are discussed correspond to the accumulation of irreversible deformation described with the theory of plasticity which is first presented. The algorithm for elasto-plasticity assuming small perturbations is then constructed. It is extended to account for geometrical effects, prior to the discussion of two simple examples. The first example concerns the strain localization in a shear band within a rectangular, compressed specimen and the second example is the buckling of an elasto-plastic plate. For sake of simplicity, the material composing these two structures has a single phase, it is a dry frictional and cohesive material. The formalism presented is readily applicable to account for the presence of a fluid phases if drained conditions are assumed.

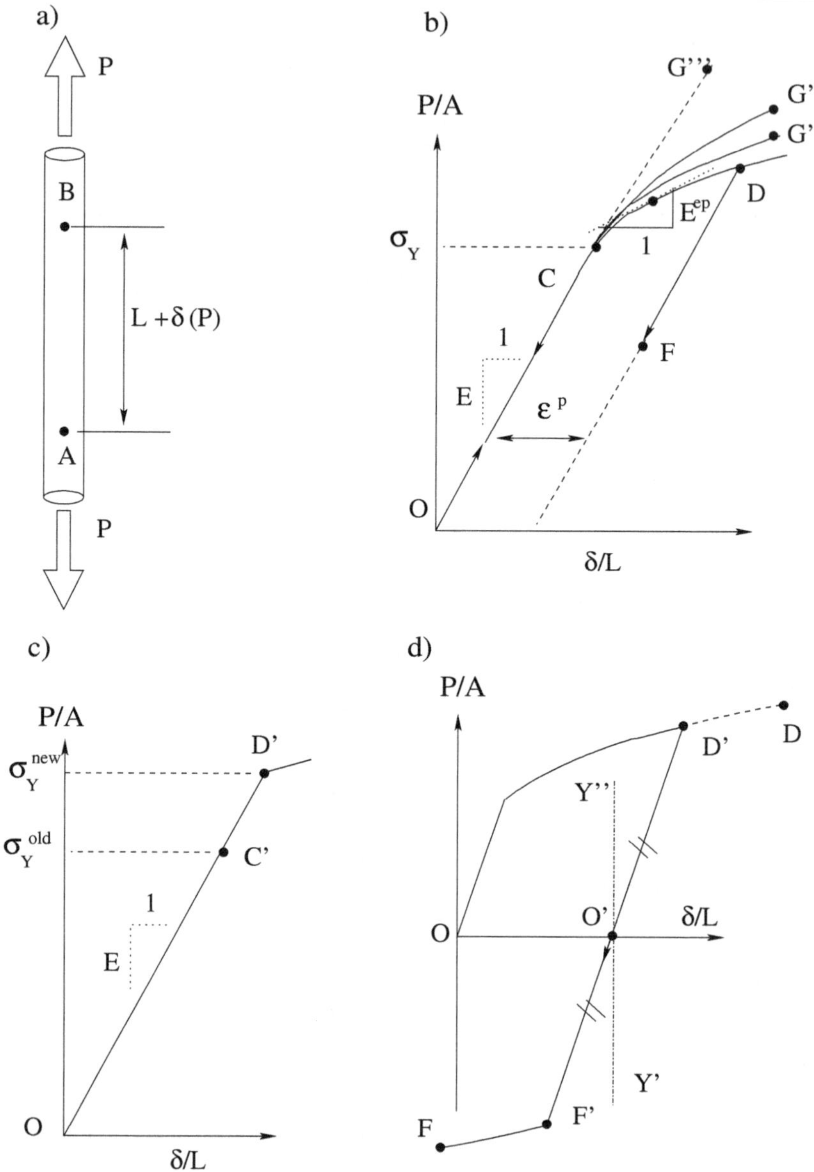

Figure 13. The idealised testing of a bar a) resulting in irreversible deformation once loaded beyond the first yield stress in tension b). A second test in c) on the same specimen shows that the material has kept memory of the first loading, with a new yield stress at the level where the applied load was stopped in the first test. The third test in d) explores the compressive stress range and illustrates the concept of isotropic hardening.

6.1 1D Theory of Plasticity and its Computational Algorithm

The reason to start with 1D experiments, its theoretical analysis and the corresponding computational algorithm, is pedagogical since most of the concepts which are introduced will be found again in the multi-axial plasticity theory presented afterwards.

The Extension of a Bar Consider the idealised experiment presented in Figure 13a, the extension of a bar, of circular cross section of area A, by the application of the tensile force of magnitude P. The distance between the two points A and B is measured continuously during the experiments and this length, initially L, varies by δ, a function of the load P. This change in length normalised by L corresponds to the nominal strain $\epsilon = \delta/L$, the abscissa of the graphs in Figure 13b to d. The vertical axis of the same graph corresponds to the applied load P divided by the area A, the 1D stress assumed homogeneous between points A and B of the bar. The stress-strain response recorded during the experiment is linear from point O to point C, Figure 13b. As the load P decreases from that point, the strain is observed to decrease following the curve obtained during the loading. This loading and unloading path is well approximated by a straight line with slope E, Young's modulus of elasticity. This linear material response ceases once the load exceeds the level of point C corresponding to the initial yield stress σ_Y. Further loading beyond that point leads to a non-linear relation recorded between point C and point D. The fact that irreversible deformation occurred is established by reducing the loading from any point between C and D. For the example of point D, the stress-strain response then does not follow the curve linking C and D but follows the straight line towards point F, which is parallel to the initial elastic response. If the unloading was pursued to bring the load to zero, there would be a residual or irreversible strain called the plastic strain ϵ^p.

One could imagine that the operator stops the test after complete unloading and gives the specimen to a colleague who is asked to test further the same cylindrical specimen in extension. The stress-strain response which is then obtained is presented in Figure 13c. The initial material response is linear with the slope E identical to the one found in the first test. The load is increased beyond the level of point C' corresponding to the level of stress $\sigma_{Y_{old}}$ which marks the onset of irreversible deformation in the first test (point C). The material response beyond that level is still linear with the same slope revealing that the material has kept memory of the previous loading during the first test. Indeed, the operator does not signal any irreversible deformation before the load reaches the point D' which corresponds to the same stress level $\sigma_{Y_{new}}$ as point D in the first test. Beyond that loading level, the material response is non-linear, following a curve which is a continuation of the one emanating from point D in Figure 13b, if the first test had been pursued. The fact that the yield stress changes with the accumulation of irreversible deformation is called work-hardening.

The next experimental observation which is discussed is the sensitivity of the results to the rate at which loading is applied, Figure 13b. The two previous experiments on a construction steel where conducted at room temperature and at a nominal strain rate of the order of $10^{-3}s^{-1}$. At that rate of loading, the activated deformation mechanisms (e.g. dislocation slip and interaction for crystalline materials) are fully developed and the material response could be said to be strain-rate independent. Repeating the same amount of loading on virgin specimens at faster nominal strain rates (up to the order $10^{+3}s^{-1}$) leads to results which are presented with two curves OG' and OG''. The material response is identical in the elastic domain. Beyond first yield, the material is stiffer for larger strain rates and the curves for faster rates are thus always above the curve OG. It is conjectured, that for very fast loading rates, and in the absence of inertia effects, the material response would be essentially linear as depicted by the line CG''' in the same graph. The physical interpretation is that the characteristic time associated to the activated, irreversible deformation mechanisms is much larger than the characteristic time of the loading. The instantaneous response of the material is thus purely elastic. This strain rate-sensitivity will be described in the next section with the theory of visco-plasticity.

The last idealised test explores the bar response under a cycle of tension and compression, Figure 13d. The first part of the test consists of an extension as done previously although the amount of extension leads to point D' which corresponds to a smaller stress than the one reached at point D in Figure 13b. The dotted curve from D' to D is thus known from the previous test. Instead of continuing the extension, an elastic unloading is considered following the line $D'O'$, the point O' being on the horizontal axis and thus the distance OO' is the residual, permanent deformation. The loading is continued in the same direction so that compressive (negative) stresses are now generated. The material response remains linear characteristic of elasticity. The first point where an irreversible deformation is detected is point F'. The segments $O'D'$ and $O'F'$ are of the same length. Plastic deformation accumulates for further compression as depicted by the curve $F'F$. The series of points between O' and F would superpose to the points between O' and D if they were flipped twice, first across the axis $Y'Y''$ and second across the horizontal axis. The plastic deformation develops identically in compression and in extension, continuing the hardening in compression which was initiated in tension. Such hardening response is said to be isotropic. Other hardening modes do exist but we will restrict ourselves to isotropic hardening in this chapter.

Plasticity Theory for 1D The objective is now to construct a theory to capture the essential properties found during the idealised experiments, limiting first ourselves to strain-rate independent plasticity.

The material response is assumed to be initially linear $\sigma = E\epsilon$ corresponding for example to the segment OC in Figure 13b. If plasticity has occurred, the material response could still be linear

$$\sigma = E(\epsilon - \epsilon^p), \tag{67}$$

describing segments such as DF in Figure 13b and obtained during unloading. This relation suggests that the total deformation can be decomposed into an elastic and a plastic part, the former being the difference between the total strain and the irreversible strain. The difficulty is to define the irreversible deformation. That deformation is zero if the stress has always been smaller than the initial yield stress σ_Y during the experiment. To characterise such stress states, we need to define the elastic domain in stress space as follows

$$E_\sigma = \{\sigma \in \mathbb{R} \quad \text{with} \quad f(\sigma, \sigma_Y) \equiv |\sigma| - \sigma_Y \leq 0\}. \tag{68}$$

It corresponds to the set of stresses for which the function f, called the yield criterion, is smaller than or equal to zero. Note that the absolute value is introduced in (68) since the same experimental results were obtained in compression, as it was revealed by the last test presented in Figure 13d. The yield function in (68) is negative or null in the elastic domain. A key feature is that such functions cannot be strictly positive for attainable stress levels. This property is certainly respected during the first part of the tensile test corresponding to the loading to point C since the yield stress σ_Y is equal to σ_C. As the stress varies between points O and C, it remains within the elastic domain and the function f is negative and increasing towards zero. The function f is zero for point C of the stress-strain curve. During loading beyond point C, the yield stress σ_Y changes as can be judged from the unloading from point D and the second test. The function f remains negative or null. This is because the yield stress, as found during the second test, has evolved from $\sigma_{Y\text{old}}$ to $\sigma_{Y\text{new}}$ corresponding to the stress level of points D and D', Figure 13b and c. For any stress below that level, the material response is elastic (f is negative).

We need to keep memory of this work-hardening defining the evolution of the yield stress during the accumulation of irreversible deformation.

This monitoring should not be based on the plastic deformation ϵ^p since it has been observed that during compression, Figure 13d, that deformation decreases whereas the evolution of the absolute value of the yield stress continues in a monotonic manner. It is thus proposed that the yield stress is a function of the accumulated plastic strain γ^p, a monotonically increasing parameter during a test. This parameter is defined by its rate $\dot{\gamma}^p = |\dot{\epsilon}^p|$. It suffices to define the hardening function $\sigma_Y(\gamma^p)$ which is extracted from the experimental data by subtracting from the recorded response the linear elastic stress-strain relation. The yield function in (68) could be thus be seen as a function of the stress and the yield stress $f(\sigma, \sigma_Y)$ or as a function of the stress and of the accumulated plastic strain $f(\sigma, \gamma^p)$. The function has the same notation f in the two instances.

The next question is how to define the onset of plasticity. It was concluded from the previous analysis that the stress level in the bar has to be identical to the current yield stress $\sigma_Y(\gamma^p)$ for plasticity to occur. However, that information is not sufficient to decide on the mode of deformation since either plasticity or elastic unloading will take place depending on the sign of the next stress increment (which is always of same sign as the stress rate $\dot{\sigma}$). For example, at point D of the first test, a negative $\dot{\sigma}$ correspond to an elastic response along line DF defining the elastic unloading. To the contrary, a positive $\dot{\sigma}$ leads to further irreversible deformation. During extension, the material response is thus summarised by

$$\begin{array}{llllll} \text{if} & \dot{\sigma} > 0 & \text{and} & \sigma = \sigma_Y(\gamma^p) & \text{then} & \dot{\epsilon}^p > 0\,, \\ \text{if} & \dot{\sigma} < 0 & \text{and} & \sigma \leq \sigma_Y(\gamma^p) & \text{then} & \dot{\epsilon}^p = 0\,. \end{array} \tag{69}$$

To generalise this conclusion for compressive loading, and also to prepare the grounds for the plasticity theory for multi-axial stress states, the following flow rule is proposed to define the plastic strain rate

$$\dot{\epsilon}^p = \dot{\gamma}^p \operatorname{sign}(\sigma)\,. \tag{70}$$

The accumulated plastic strain rate $\dot{\gamma}^p = |\dot{\epsilon}^p|$ defines the magnitude of the plastic strain rate which sign is given by the sign of the stress.

The equivalent plastic strain rate $\dot{\gamma}^p$ is now computed directly by inspection of the yield criterion. It has been established that $f(\sigma, \gamma^p)$ is always identical to zero during plastic straining. If the sign of the stress rate does not change then $\dot{f}(\sigma, \gamma^p) = 0$ must be satisfied during the whole plastic straining, a condition referred to as the consistency condition on the yield criterion. From the definition of the yield criterion in (68), the plastic strain rate is found from

$$\dot{f}(\sigma, \gamma^p) \equiv \operatorname{sign}(\sigma)\dot{\sigma} - \sigma_Y' \dot{\gamma}^p = 0 \Rightarrow \dot{\gamma}^p = \frac{\operatorname{sign}(\sigma)\dot{\sigma}}{\sigma_Y'}\,, \tag{71}$$

in which σ'_Y is the derivative of the yield stress with respect to the accumulated plastic strain.

To complete this 1D, strain-rate independent plasticity theory, it is worth to construct the relation between the rates of stress and strain assuming plasticity is taking place. The relation

$$\dot{\sigma} = E(\dot{\epsilon} - \dot{\epsilon}^p) , \tag{72}$$

obtained from (67) is the starting point. The plastic strain rate is expressed in terms of the stress rate thanks to the flow rule (70) and the consistency condition (71) holds so equation (72) becomes

$$\dot{\sigma} = E^{ep}\dot{\epsilon} \quad \text{with} \quad E^{ep} = \frac{E\sigma'_Y}{E + \sigma'_Y} , \tag{73}$$

in terms of the elasto-plastic tangent E^{ep}, shown at an arbitrary point between points C and D in Figure 13b.

The final word on this 1D theory concerns the extension to account for strain-rate effects discussed in the previous subsection. The definition (68) of the elastic domain still holds. The stress has to reach $\pm\sigma_Y$ to initiate plasticity. The difference is that we accept that this yield stress could be exceeded during loading to reflect the idea that the deformation mechanisms do not have time to be fully activated. The equivalent plastic strain rate is then defined in terms of the distance between the stress and the yield stress, the simplest relation being the linear overstress model

$$\dot{\gamma}^p = \frac{< |\sigma| - \sigma_Y(\gamma^p) >}{\eta} , \tag{74}$$

in which η is the viscosity (unit Pas) and the brackets $< a >$ defines the function equal to a if a is positive and equal to zero otherwise. The consistency condition (71) is thus replaced by an ordinary differential equation (74). Plasticity occurs as long as $|\sigma|$ is larger than $\sigma_Y(\gamma^p)$. Consequently, there is plastic flow during the initial part of an unloading that would start after loading to point G', Figure 13b. Note that if the strain rate is tending to zero then (74) provides the condition $|\sigma| = \sigma_Y(\gamma^p)$ which is the yield criterion considered for the strain-rate independent theory above.

Update Algorithms, Example of the Backward Euler Scheme Time is now discretised and it is assumed that at time t_n, the strain ϵ_n, the accumulated plastic strain γ^p_n and the stress σ_n are known. A displacement-based finite-element approximation provides candidates for the total strain at time t_{n+1}, denoted ϵ_{n+1}, as it will be explained in the next subsection.

The objective of the update algorithm is to determine the new value of the accumulated plastic strain γ^p_{n+1} and the stress σ_{n+1} at the end of the time step ($t_{n+1} = t_n + \Delta t$). Note that the subscript defines the moment of time at which the variable is evaluated.

The update algorithm is classically obtained from the rate equations. For example, if the governing equation, with initial conditions based on the beginning of the increment, is

$$\dot{x}(t) = \mathcal{F}(x,t)\,, \quad x(t_n) = x_n\,, \tag{75}$$

then the algorithm will be

$$x_{n+1} = x_n + \Delta t \mathcal{F}(x_{n+\beta}, t_{n+1}) \quad \text{with} \quad x_{n+\beta} = \beta x_{n+1} + (1-\beta)x_n\,, \tag{76}$$

in which the scalar β is within $[0,1]$ and the scalar $x_{n+\beta}$ is a convex combination of x_n and x_{n+1}. The algorithm is called forward Euler, the mid-point rule, and backward Euler for the scalar β set to 0, 1/2 and 1, respectively. The algorithm is explicit only for the forward Euler scheme since the right-hand side of (76a) is then independent of the value of the unknown x_{n+1} at the end of the time step. The two other schemes are said to be implicit. For reasons of stability and accuracy, the implicit schemes are often favoured. It is the backward Euler scheme which is now explored in details for the 1D plasticity theory presented above.

The rate equations for the strain-rate independent plasticity theory, assuming a single internal variable which is the accumulated plastic strain, becomes after time discretisation

$$\begin{aligned} \sigma_{n+1} &= \sigma_n + E\Big(\Delta\epsilon - \Delta\gamma^p \operatorname{sign}(\sigma_{n+1})\Big) \\ f(\sigma_{n+1}, \gamma^p_{n+1}) &\equiv \sigma^e_{n+1} - \sigma_Y(\gamma^p_{n+1}) \leq 0\,, \end{aligned} \tag{77}$$

in which the yield criterion is satisfied at the end of the increment and the equivalent stress σ^e is the absolute value of the stress $|\sigma|$. All the quantities at time t_n are known and the increment in strain is also assumed to be given. The solution of this update algorithm, for the quantities at time t_{n+1} is constructed in two steps, referred to as the predictor and the corrector step. For the predictor step, one assumes that there is no plasticity upon further straining by $\Delta\epsilon$. The predictor stress and plastic strain are thus based on the elastic incremental relation

$$\sigma_* = \sigma_n + E\Delta\epsilon \quad , \gamma^p_* = \gamma^p_n\,, \tag{78}$$

in which a star in subscript identifies quantities attached to the predictor. The question is then whether this predictor stress satisfies the yield condition $f(\sigma_*, \gamma^p_*) \leq 0$. If so, no plastic flow occurs, the increment is indeed purely elastic and the final state is defined by the predictor

$$\sigma_{n+1} = \sigma_* \quad , \gamma^p_{n+1} = \gamma^p_* \,. \tag{79}$$

If the yield condition based on the predictor is not satisfied, one needs, in the second step, to correct the stress to account for the plastic straining according to

$$\sigma_{n+1} = \sigma_* - \Delta\gamma^p E \;\mathrm{sign}(\sigma_{n+1}) \quad , \text{ with } \quad \mathrm{sign}(\sigma_{n+1}) = \frac{\sigma_{n+1}}{\sigma^e_{n+1}} \,. \tag{80}$$

This correction is now presented in a manner which is awkward for this 1D setting but which prepares the grounds for the construction of the multi-dimensional stress update algorithm. Two informations are now extracted from (80). First, group terms in the same side of (80) which are related to time t_{n+1}

$$\sigma_{n+1}\Big(1 + \frac{\Delta\gamma^p E}{\sigma^e_{n+1}}\Big) = \sigma_* \,.$$

Second, take the absolute value on the two sides of this last equation to obtain the relation between σ^e_{n+1} and the equivalent predictor stress σ^e_* and the accumulated plastic strain increment. These two relations are summarised by

$$\begin{aligned} \sigma_{n+1} &= \sigma_* \{1 + \frac{\Delta\gamma^p E}{\sigma^e_{n+1}}\}^{-1} \,, \\ \sigma^e_{n+1} &= \sigma^e_* - E\Delta\gamma^p \,. \end{aligned} \tag{81}$$

In (81a), the final stress results from the division of the stress predictor by a scalar, the same division will appear in the algorithm for multi-axial stress state. Note that the two stresses σ_{n+1} and σ^e_{n+1} are only function of the predictor stress (already computed) and the accumulated plastic strain increment, the only remaining unknown. To find this remaining unknown $\Delta\gamma^p$, one enforces the yield condition at the end of the time step and make use of (81b) to obtain

$$f(\sigma^e_* - E\Delta\gamma^p, \gamma^p_n + \Delta\gamma) \equiv \sigma^e_* - E\Delta\gamma^p - \sigma_Y(\gamma^p_n + \Delta\gamma^p) = 0 \,. \tag{82}$$

This equation has a single unknown, the increment in accumulated plastic strain. It is most often non-linear as suggested by the work-hardening in

Figure 13. The solution of (82) is then found by iterative search (Newton-Raphson). Once $\Delta\gamma^p$ is determined, the update algorithm is continued by computing the equivalent stress σ^e_{n+1} (81b) and finalised by the division in (81a) for the stress σ_{n+1}.

Note that for visco-plasticity, the algorithm is identical except that the yield condition (77b) is replaced by the viscosity function which, for the linear overstress model proposed in (74), reads

$$\Delta\gamma^p = \frac{\Delta t}{\eta}\Big(\sigma^e_{n+1} - \sigma_Y(\gamma^p_n + \Delta\gamma^p)\Big) . \tag{83}$$

The equivalent stress is expressed in terms of the stress predictor and $\Delta\gamma^p$ and (83) constitutes a single non-linear equation for that unknown. Once determined, the rest of the update is identical for rate-dependent and rate-independent plasticity.

Linearisation and Consistent Tangent The starting point of the finite-element formulation is not the total energy of the system, as for the previous sections, because of the possibility of unloading. The finite-element is based simply on the weak from of equilibrium provided by the theorem of virtual work (13) which 1D version is

$$\int_\Omega \sigma\hat{\epsilon} dV - \int_\Omega \rho g \hat{u} dV - \int_{\partial\Omega} T^d \hat{u} dS = 0 , \tag{84}$$

for a domain Ω of boundary $\partial\Omega$ and any K.A. displacement field $\hat{u}$ ($\hat{u}$ is continuous and equal to zero on $\partial\Omega_u$). Recall that $\hat{\epsilon}$ is computed from the K.A. displacement field. Discretise in time and enforce equilibrium at the end of the time step so that (84) becomes

$$\int_\Omega \sigma_{n+1}\hat{\epsilon} dV - \int_\Omega \rho g \hat{u} dV - \int_{\partial\Omega} T^d_{n+1} \hat{u} dS = 0 . \tag{85}$$

The stress σ_{n+1} is constructed with the non-linear update algorithm discussed above and is thus a function of the stress and the accumulated plastic strain estimated at the beginning of the time step as well as of the increment in strain. This update algorithm will be denoted $\sigma_{n+1} \equiv \Sigma(\sigma_n, \gamma^p_n, \Delta\epsilon)$ in which Σ is a stress function of σ_n, γ^p_n and $\Delta\epsilon$. Equation (85) cannot be solved directly because of the material non-linearities present in the update algorithm. An iterative scheme, based on Newton-Raphson, is often proposed to find the solution in terms of displacement. A series of strain increments is constructed $\Delta\epsilon^{(k)}$ and the corresponding series in stress is obtained with $\sigma^{(k)}_{n+1} \equiv \Sigma(\sigma_n, \gamma^p_n, \Delta\epsilon^{(k)})$ with initial conditions $\Delta\epsilon^{(0)} = 0$ and

$\sigma_{n+1}^{(0)} = \sigma_n$. The iteration counter k is in upperscript. To proceed from iteration k to the next, one linearises the update algorithm

$$\sigma_{n+1}^{(k+1)} \simeq \sigma_{n+1}^{(k)} + \frac{\partial \Sigma}{\partial \Delta\epsilon}(\sigma_n, \gamma_n^p, \Delta\epsilon^{(k)})\, \tilde{\epsilon}^{(k+1)} \quad \text{with} \quad \Delta\epsilon^{(k)} = \sum_{j=0}^{k} \tilde{\epsilon}^{(j)}\,, \tag{86}$$

in which $\tilde{\epsilon}^{(k+1)}$ is the correction to the strain increment at iteration $k+1$. Insert the development (86) in equation (85) to obtain

$$\begin{aligned} &\int_\Omega \hat{\epsilon} \frac{\partial \Sigma}{\partial \Delta\epsilon}(\sigma_n, \gamma_n^p, \Delta\epsilon^{(k)})\, \tilde{\epsilon}^{(k+1)} dV = \\ &\int_\Omega \rho g \hat{u}\, dV + \int_{\partial\Omega} T_{n+1}^d \hat{u}\, dS - \int_\Omega \hat{\epsilon} \sigma_{n+1}^{(k)} dV\,. \end{aligned} \tag{87}$$

The right-hand side of (87) is based on quantities known at the end of iteration k and defines the out-of-balance force, the difference between the internal and the external work. On the left-hand side, one finds the stiffness matrix and the only unknown of the iteration $\tilde{\epsilon}^{(k+1)}$. It remains to express the strain unknown in terms of the nodal displacement iterations, with the introduction of the finite-element approximation, following the same approach as for linear problems (see section 3). Convergence is achieved once the norm of the out-of-balance force, defined by the norm of the right-hand side of (87) is sufficiently small: the internal work is then equal to the external work as it should be in equilibrium. It is of note that the tangent stiffness in (87) has been constructed with the linearisation of the non-linear update algorithm, and not with the elasto-plastic tangent defined in (73). This choice is fundamental to guarantee the quadratic convergence of the Newton-Raphson equilibrium search (Simo and Taylor, 1985). It remains to construct explicitly this consistent tangent $\partial\Sigma/\partial\Delta\epsilon$ for the implicit update discussed earlier. This task is postponed to the next section on multi-axial stress states since the 1D setting is not prompt to illustrate the concept.

The final remark on the finite-element method for non-linear materials is that there are two non-linear solution searches solved by iterative means (Newton-Raphson). The first iteration is local, at the quadrature point, and is due to the material non-linearities which are treated in the non-linear update algorithm. The second iterative process is global, for the nodal displacement, and is due to the equilibrium search which is enforced at the beginning and the end of the time step.

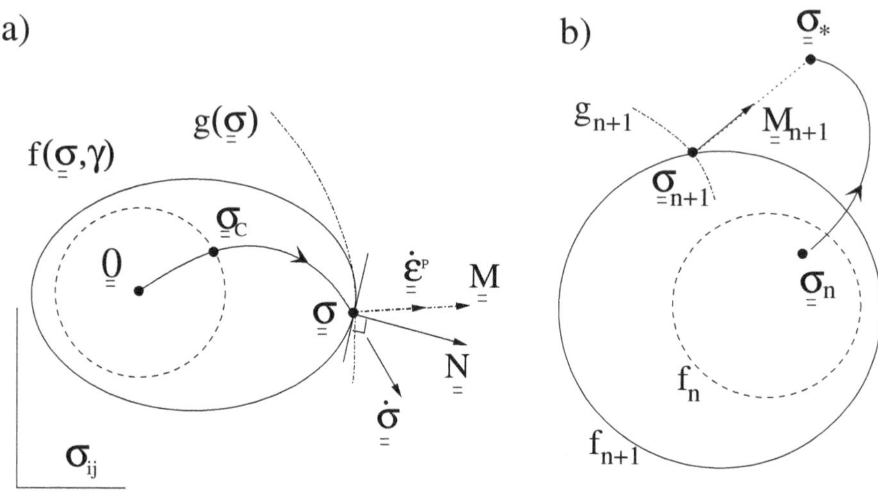

Figure 14. The yield surface $f(\underline{\underline{\sigma}}, \gamma^p)$ and the flow potential $g(\underline{\underline{\sigma}})$ in the six-dimensional stress space in a). The normal to the yield surface $\underline{\underline{N}}$ is different from the normal to the flow potential $\underline{\underline{M}}$ for non-associated plasticity. The decomposition of the update in the computation of a predictor stress $\underline{\underline{\sigma}}_*$, based on the material elastic response and a plastic correction, along the direction of the normal to the flow potential at the end of the time step, is shown in b).

6.2 Extension to Multi-Axial Stress States, Assuming Small Perturbations

The objective is now to generalize the 1D plasticity theory and its numerical algorithm to multi-dimensional stress states. A few words on the theory is first proposed, followed by the update algorithm and finally, the derivation of the consistent tangent defined above.

Plasticity Theory The symmetric stress tensor has six independent components and the elastic domain is thus now defined in the six-dimensional stress space, Figure 14. In the geometrical construction proposed in this Figure, the tensors are represented as classical vectors as if they were plotted in our 3D physical space. In particular, the double dot product $\underline{\underline{A}} : \underline{\underline{B}}$ between two tensors is equal to the scalar resulting from the classical dot or scalar product $\underline{a} \cdot \underline{b}$ between the two vectors $\underline{a}$ and $\underline{b}$ representing graphically the tensors $\underline{\underline{A}}$ and $\underline{\underline{B}}$, respectively. This analogy will be useful to understand various steps of the reasoning to be followed. The elastic domain in this six-dimensional space corresponds to the set of stress tensors

$$E_\sigma = \{\underline{\underline{\sigma}} \in \mathrm{I\!R}^6 \quad \text{with} \quad f(\underline{\underline{\sigma}}, \gamma^p) \le 0\}\,. \tag{88}$$

It is assumed, again for simplicity sake, that a single internal variable, the accumulated plastic strain γ^p, describes the evolution of the yield surface due to work-hardening. This evolution is illustrated in Figure 14a where the first yield occurs at point $\underline{\underline{\sigma}}_C$ on the dashed surface. The shape of the yield surface is convex and has often symmetries in the stress space. If yield is isotropic, the elastic domain boundary is described with the three stress invariants (Hill, 1952). For example, the elastic domain of most metals is usually described using the square root of the second invariant J_2 of the deviatory stress ($\underline{\underline{\sigma}}'$). In summary, for most metals

$$\begin{gathered}
\sigma^e_{VM} = \sqrt{J_2} \quad \text{with } J_2 = \frac{1}{2}\underline{\underline{\sigma}}' : \underline{\underline{\sigma}}' \quad \text{and} \quad \underline{\underline{\sigma}}' = \underline{\underline{\sigma}} - \frac{1}{3}\underline{\underline{\delta}}(\underline{\underline{\delta}} : \underline{\underline{\sigma}})\,, \\
\left(J_2 = \frac{1}{2}\sigma'_{ij}\sigma'_{ji} \quad , \quad \sigma'_{ij} = \sigma_{ij} - \frac{1}{3}\delta_{ij}\sigma_{kk}\right), \\
f(\sigma^e_{VM}, \gamma^p) = \sigma^e_{VM} - \sigma_Y(\gamma^p) \le 0\,.
\end{gathered} \tag{89}$$

For this J_2 or Von Mises plasticity model, the yield surface is described by $f(\sigma^e_{VM}, \gamma^p)$ and is independent of the first stress invariant (or the hydrostatic stress P). Note that the factor 1/2 in the definition of J_2 ensures that the equivalent stress collapses to the 1D stress theory discussed above for a simple shear test. A second example of interest is the Drucker-Prager criterion for which the equivalent stress is

$$\begin{gathered}
\sigma^e_{DP} = \sqrt{J_2} + \mu P \quad \text{with} \quad P = \frac{1}{3}(\underline{\underline{\delta}} : \underline{\underline{\sigma}})\,, \\
f(\sigma^e_{DP}, \gamma^p) = \sigma^e_{DP} - \sigma_Y(\gamma^p) \le 0\,,
\end{gathered} \tag{90}$$

in which μ and P are the friction coefficient and the hydrostatic stress, respectively. The yield stress σ_Y receives the name of cohesion in rock and soil plasticity and will be denoted c in what follows.

Plastic flow takes place if the stress point in on the boundary of the elastic domain, the yield surface, and if the stress rate is pointing outside the elastic stress domain, Figure 14a. This condition is met at the current stress point $\underline{\underline{\sigma}}$, which is on the yield surface, if the stress rate $\dot{\underline{\underline{\sigma}}}$ is pointing towards the sub-space bounded by the plane tangent to the stress point on the yield surface and pointed by the normal tensor $\underline{\underline{N}}$. The scalar $\dot{\underline{\underline{\sigma}}} : \underline{\underline{N}}$ corresponds to the scalar product of the two vectors presented graphically. Plastic flow requires this scalar to be positive. The loading conditions at $f(\underline{\underline{\sigma}}, \gamma^p) = 0$ are thus summarised by

$$\begin{aligned} \dot{\underline{\underline{\sigma}}} : \underline{\underline{N}} > 0 &\qquad \text{plastic flow}\,, \\ \dot{\underline{\underline{\sigma}}} : \underline{\underline{N}} < 0 &\qquad \text{elastic unloading}\,, \\ \dot{\underline{\underline{\sigma}}} : \underline{\underline{N}} = 0 &\qquad \text{neutral loading}\,. \end{aligned} \tag{91}$$

The normal to the yield surface for the criterion (89) and (90) reads

$$\begin{aligned} \text{Von Mises:} \quad \underline{\underline{N}} &= \frac{1}{2\sqrt{J_2}}\underline{\underline{\sigma}}'\,, \\ \text{Drucker-Prager:} \quad \underline{\underline{N}} &= \frac{1}{2\sqrt{J_2}}\underline{\underline{\sigma}}' + \mu\frac{1}{3}\underline{\underline{\delta}}\,. \end{aligned} \tag{92}$$

If plastic flow is detected, we need to define the plastic flow directions in the stress space. This task is most often done with the introduction of a second potential, the flow potential g, function of the stress state. The flow rule defines the direction of the plastic strain rate to be along the normal $\underline{\underline{M}}$ to the flow potential in stress space

$$\dot{\underline{\underline{\epsilon}}}^p = \dot{\gamma}^p \underline{\underline{M}} \quad \text{with} \quad \underline{\underline{M}} = \frac{\partial g}{\partial \underline{\underline{\sigma}}} \quad \left(M_{ij} = \frac{\partial g}{\partial \sigma_{ij}} \right). \tag{93}$$

The directions of the plastic strain rate are thus independent of the stress rate tensor and are fully determined by the current stress only. For most metals, the flow potential is again Von Mises equivalent stress (89): $g(\underline{\underline{\sigma}}) = \sigma^e_{VM}$. For rocks and soils, the flow potential is usually dependent on the first stress invariant. The simplest relation is of the same type as in (90) and is written as

$$g(\underline{\underline{\sigma}}) = \sqrt{J_2} + \beta P\,, \tag{94}$$

with the introduction of the scalar β, assumed constant for sake of simplicity, which deserves some discussion. The trace of the plastic strain rate in (93), $\mathrm{tr}(\dot{\underline{\underline{\epsilon}}}^p) = \dot{\gamma}^p \,\mathrm{tr}(\underline{\underline{M}})$, is the rate of irreversible volume change. For most metals, the plastic volume change is negligible explaining the selection of a flow potential independent of the first stress invariant and thus $\mathrm{tr}(\underline{\underline{M}}) = 0$ so that indeed $\mathrm{tr}(\dot{\underline{\underline{\epsilon}}}^p) = 0$. For rocks and soils, this is certainly not the case during the early part of the plastic deformation. The trace of the plastic strain rate is $\dot{\gamma}^p\beta$ by combining (93) and (94). This observation explains that the scalar β defines the material irreversible dilatancy upon plastic deformation. It is certainly a complex function of the stress state and of the degree of work-hardening for most soils.

Note that if the two normals $\underline{\underline{M}}$ and $\underline{\underline{N}}$ are identical (same potential $f = g$), the plasticity model is said to be associated and it is non-associated otherwise. Frictional and cohesive materials usually require a non-associated description with consequences on the symmetry of the elasto-plastic tangent, as it will be seen next. Prior to this derivation, we need to define the amplitude of the plastic strain rate $\dot{\gamma}^p$. The starting point is the consistency criterion $\dot{f} = 0$ during continuous plastic straining which provides

$$\dot{f}(\underline{\underline{\sigma}}, \gamma^p) = 0 \Rightarrow \underline{\underline{N}} : \dot{\underline{\underline{\sigma}}} + f_{,\gamma^p}\dot{\gamma}^p = 0 \,. \tag{95}$$

The strain rate is still assumed to be decomposed in an elastic and a plastic part. The stress rate is thus linearly related to the elastic strain rate with

$$\dot{\underline{\underline{\sigma}}} = \mathbb{D}^e : (\dot{\underline{\underline{\epsilon}}} - \dot{\underline{\underline{\epsilon}}}^p) \,, \tag{96}$$

in which the elasticity tensor $\mathbb{D}^e$ is defined in (4) in the case of isotropy. Combine (96) with the consistency condition (95) and express the rate of accumulated plastic strain in terms of the strain rate

$$\dot{\gamma}^p = \frac{1}{H}\underline{\underline{N}} : \mathbb{D}^e : \dot{\underline{\underline{\epsilon}}} \quad \text{with} \quad H = \underline{\underline{N}} : \mathbb{D}^e : \underline{\underline{M}} - f_{,\gamma^p} \,, \tag{97}$$

$$\left(\dot{\gamma}^p = \frac{1}{H} N_{ij}\ \mathbb{D}^e_{jikl}\dot{\epsilon}_{lk} \quad , \qquad H = N_{ij}\ \mathbb{D}^e_{jikl} M_{lk} - f_{,\gamma^p}\right).$$

This result and the flow rule (93) are finally combined in the rate equation (96) to provide

$$\dot{\underline{\underline{\sigma}}} = \mathbb{D}^{ep} : \dot{\underline{\underline{\epsilon}}} \quad \text{with} \quad \mathbb{D}^{ep} = \mathbb{D}^e - \frac{1}{H}\ \mathbb{D}^e : \underline{\underline{M}} \otimes \underline{\underline{N}} : \mathbb{D}^e \,, \tag{98}$$

$$\left(\ \mathbb{D}^{ep}_{jikl} = \mathbb{D}^e_{jikl} - \frac{1}{H}\ \mathbb{D}^e_{jipq} M_{qp} N_{mn}\ \mathbb{D}^e_{nmkl}\right),$$

in which $\mathbb{D}^{ep}$ is the plasticity tangent operator. For a frictional material with the Drucker-Prager yield criterion (89) and a zero dilatancy ($\beta = 0$ in 94), the plasticity tangent tensor is

$$\mathbb{D}^{ep} = \mathbb{D}^e - \frac{2G}{H}[2G\underline{\underline{M}} \otimes \underline{\underline{M}} + \mu K \underline{\underline{M}} \otimes \underline{\underline{\delta}}] \,, \tag{99}$$

in which G and K are the shear and bulk elasticity modulus and μ the friction coefficient.

Update Algorithms An implicit update algorithm is now presented for non-associated plasticity assuming a single internal variable (the accumulated plastic strain γ^p), no plastic dilatancy and an isotropic elastic medium. The rate equations (96) are replaced with the incremental relations

$$\underline{\underline{\sigma}}_{n+1} = \underline{\underline{\sigma}}_n + \mathbb{D}^e : (\Delta\underline{\underline{\epsilon}} - \Delta\gamma^p \underline{\underline{M}}_{n+1}) , \qquad (100)$$

$$\text{with} \quad \underline{\underline{M}}_{n+1} = \frac{\partial g}{\partial \sigma}|_{n+1} \quad \text{and} \quad g(\underline{\underline{\sigma}}_{n+1}) = \sqrt{J_2}_{n+1} ,$$

which make use of the flow rule defined in (93), estimated at the end of the increment. The stress at the end of the increment appears on both side of the equation (100) and the update is thus implicit. The yield condition, chosen to be the Drucker-Prager criterion, is enforced at the end of the time step

$$f(\underline{\underline{\sigma}}_{n+1}, \gamma^p_{n+1}) \equiv \sqrt{J_2}_{n+1} + \mu P_{n+1} - c(\gamma^p_{n+1}) \leq 0 , \qquad (101)$$

in which c is the cohesion, the function of the accumulated plastic strain representing the work-hardening. The stress and internal variable are known at time t_n and a strain increment $\Delta\underline{\underline{\epsilon}}$ is proposed. The objective is to determine the stress $\underline{\underline{\sigma}}_{n+1}$ and γ^p_{n+1} at the end of the time step, generalizing the scheme presented in section 6.1.

The solution search is decomposed into two steps, the first being the predictor, defined as if further straining is in the elastic range of deformation

$$\underline{\underline{\sigma}}_* = \underline{\underline{\sigma}}_n + \mathbb{D}^e : \Delta\underline{\underline{\epsilon}} \quad , \quad \gamma^p_* = \gamma^p_n , \qquad (102)$$

corresponding to (78) in 1D. If the yield condition $f(\underline{\underline{\sigma}}_*, \gamma^p_*) \leq 0$ is respected, the increment is indeed elastic and the predictor is the final stress at the end of the increment. If the yield condition is not respected, the second step of the algorithm – the plastic corrector – has to be considered

$$\underline{\underline{\sigma}}_{n+1} = \underline{\underline{\sigma}}_* - \Delta\gamma^p 2G \underline{\underline{M}}_{n+1} \quad \text{with} \quad \underline{\underline{M}}_{n+1} = \Big(\frac{1}{2\sqrt{J_2}} \underline{\underline{\sigma}}'\Big)|_{n+1} , \qquad (103)$$

$$\sqrt{J_2}_{n+1} + \mu P_{n+1} - c(\gamma^p_n + \Delta\gamma^p) = 0 .$$

The solution of this set of equations is sought by decomposing the stress in a spherical and deviatory part. Equations (103a) then provide

$$P_{n+1} = P_* \quad , \quad \underline{\underline{\sigma}}'_{n+1}(1 + \frac{G\Delta\gamma^p}{\sqrt{J_2}_{n+1}}) = \underline{\underline{\sigma}}'_* . \qquad (104)$$

The second invariant of the deviatory stress is readily found from (104b) by taking the double dot product of each member of this equation by itself (e.g. $\underline{\underline{\sigma}}'_* : \underline{\underline{\sigma}}'_*$ for the right-hand side), leading to

$$\sqrt{J_{2_{n+1}}} + G\Delta\gamma^p = \sqrt{J_{2_*}}\,. \tag{105}$$

The two invariants of the stress tensor at time t_{n+1} in (104a) and (105) are thus expressed in terms of the two invariants of the predictor stress so that the yield condition (103b) now reads

$$\sqrt{J_{2_*}} + \mu P_* - G\Delta\gamma^p - c(\gamma_n^p + \Delta\gamma^p) = 0\,, \tag{106}$$

in terms of a single variable, the increment of equivalent plastic strain $\Delta\gamma^p$. This equation could be non-linear if the cohesion is a non-linear function of that variable and is then solved by iterative means (Newton-Raphson). Once $\Delta\gamma^p$ is determined the stress deviator is then found from (104b) by scaling the deviatory predictor stress with a single scalar, a procedure often called the radial return because of its first discussion in the case of Von Mises associated plasticity.

Consistent Tangent Equilibrium is enforced at the end of the time step with the application of the theorem of virtual work

$$\int_\Omega \underline{\underline{\sigma}}_{n+1} : \hat{\underline{\underline{\epsilon}}}\, dV - \int_\Omega \rho \underline{g} \cdot \hat{\underline{u}}\, dV - \int_{\partial\Omega} \underline{T}^d_{n+1} \cdot \hat{\underline{u}}\, dS = 0\,, \tag{107}$$

generalizing (85) to multi-axial stress states. The stress $\underline{\underline{\sigma}}_{n+1}$ is defined by the non-linear update algorithm denoted now by the tensor $\underline{\underline{\Sigma}}(\underline{\underline{\sigma}}_n, \gamma_n^p, \Delta\underline{\underline{\epsilon}})$ and function of the stress and internal variable at time t_n and of the increment in strain. This non-linear equation is solved by Newton-Raphson and the linearisation at the iteration $k+1$ of the increment provides

$$\int_\Omega \hat{\underline{\underline{\epsilon}}} : \frac{\partial \underline{\underline{\Sigma}}}{\partial \Delta \underline{\underline{\epsilon}}}|^{(k)} : \tilde{\underline{\underline{\epsilon}}}^{(k+1)}\, dV = \tag{108}$$
$$\int_\Omega \rho \underline{b} \cdot \hat{\underline{u}}\, dV + \int_{\partial\Omega} \underline{T}^d_{n+1} \hat{\underline{u}}\, dS - \int_\Omega \hat{\underline{\underline{\epsilon}}} : \underline{\underline{\sigma}}^{(k)}_{n+1}\, dV\,.$$

The objective of this section is to construct the consistent tangent $\partial\underline{\underline{\Sigma}}/\partial\Delta\underline{\underline{\epsilon}}$ from the linearisation of the implicit update algorithm presented above.

The linearised algorithm is summarised by

$$\begin{gathered}\tilde{\underline{\underline{\sigma}}}'_* = 2G\tilde{\underline{\underline{\epsilon}}}' \quad , \quad \tilde{P}_* = K\underline{\underline{\delta}} : \tilde{\underline{\underline{\epsilon}}} \quad , \sqrt{\tilde{J_2}}_* = 2G\underline{\underline{M}} : \tilde{\underline{\underline{\epsilon}}} , \\ \sqrt{\tilde{J_2}} = \sqrt{\tilde{J_2}}_* - G\tilde{\gamma}^p , \\ \sqrt{\tilde{J_2}}_* + \mu\tilde{P}_* - (G + c_{,\gamma^p})\tilde{\gamma}^p = 0 , \\ \tilde{\underline{\underline{M}}} = -\frac{\sqrt{\tilde{J_2}}_*}{2J_{2*}}\underline{\underline{\sigma}}'_* + \frac{1}{2\sqrt{J_2}_*}\tilde{\underline{\underline{\sigma}}}'_* , \\ \tilde{\underline{\underline{\sigma}}}' = \tilde{\underline{\underline{\sigma}}}'_* - \tilde{\gamma}^p 2G\underline{\underline{M}} - \Delta\gamma^p 2G\tilde{\underline{\underline{M}}} \quad , \tilde{P} = \tilde{P}_* ,\end{gathered} \tag{109}$$

in which, again, the superposed ˜ identifies the correction at the $k+1$ iteration and replaces this counter for sake of clarity. Note that all variables are estimated at the iteration k of the current time increment leading to t_{n+1}, which is not stated in the set of equations (109) also for sake of clarity. The first line of equations in (109a) results from the linearisation of the stress predictor (102) and its invariants. Equations (109b) and (109c) are directly the linearisation of the relation between the two stress invariants in (105) and of the yield criterion (106), which is assumed to be satisfied at the next iteration on the increment. Equations (109d) and (109e) correspond to the linearisation of the flow rule and of the correction step found in (104a). Note that the linearisation of the flow rule is not based on the final stress but on the predictor stress, this substitution being permitted since the ratios $\underline{\underline{\sigma}}'/\sqrt{J_2}$ and $\underline{\underline{\sigma}}'_*/\sqrt{J_2}_*$ are identical according to (103b). This observation allows us to express directly the linearisation of the flow direction in terms of the linearised stress predictor and thus of the linearised, deviatory strain using (109a)

$$\tilde{\underline{\underline{M}}} = \frac{G}{2\sqrt{J_2}_*}[\, \mathbb{I}_S - \underline{\underline{M}} \otimes \underline{\underline{M}}] : \tilde{\underline{\underline{\epsilon}}}' , \tag{110}$$

in which $\mathbb{I}_S$ is the symmetric fourth-order identity tensors introduced already in (4). The system of equations in (109) is solved by first relating the iteration $\tilde{\gamma}^p$ from (109c) to the strain tensor with the help of (109a)

$$\tilde{\gamma}^p = \frac{1}{H}(\underline{\underline{M}} + \mu K\underline{\underline{\delta}}) : \tilde{\underline{\underline{\epsilon}}} \quad \text{with} \quad H = G + c_{,\gamma^p} . \tag{111}$$

The reader is then ready to combine these two preliminary results (110) and (111) in (109e) and to include the linearised spherical stress to obtain the consistent tangent

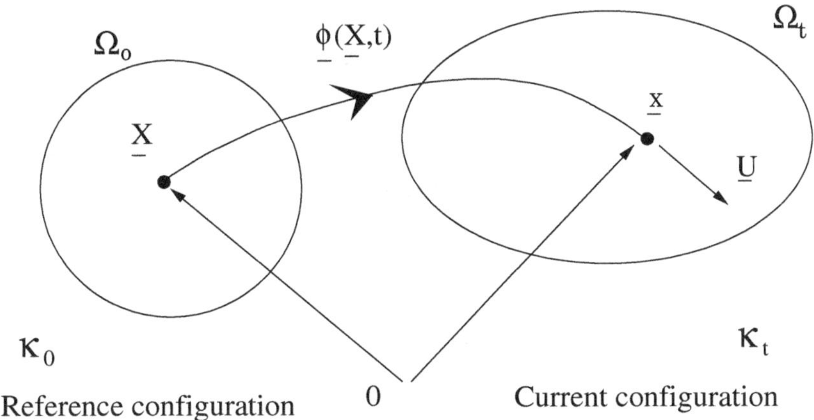

Figure 15. The continuum occupies initially the domain Ω_o in the reference configuration κ_0 and is convected to occupy the domain Ω_t in the current configuration κ_t. The transformation is described by the mapping $\underline{\phi}(\underline{x}, t)$.

$$\begin{aligned}\frac{\partial \underline{\underline{\Sigma}}}{\partial \Delta \underline{\underline{\epsilon}}} &= \mathbb{D}^e - \frac{2G}{H}[2G\underline{\underline{M}} \otimes \underline{\underline{M}} + \mu K \underline{\underline{M}} \otimes \underline{\underline{\delta}}] \\ &\quad - \Delta\gamma^p \frac{2G^2}{\sqrt{J_{2*}}}[\,\mathbb{I}_S - \frac{1}{3}\underline{\underline{\delta}} \otimes \underline{\underline{\delta}} - \underline{\underline{M}} \otimes \underline{\underline{M}}]\,.\end{aligned} \tag{112}$$

At the first iteration ($\Delta\gamma^p = 0$) of the increment, the last term is zero and the consistent tangent is identical to the elasto-plastic tangent operator in (99) for this specific case of an isotropic elastic medium, the Drucker Prager yield criterion and the selection of a non-dilatant plastic flow rule. During the iterative process, the consistent tangent differs from the elasto-plastic tangent operator because of the last term in (112) which is due to the linearisation of the flow rule. This additional term is important to ensure a quadratic rate of convergence of the equilibrium search.

6.3 Extension to Finite-Transformation

The last technical difficulty of this chapter is the account of finite transformations. It is necessary to review first a few concepts of continuum mechanics, exposed in details in reference books such as Malvern (1969), Salençon (2001) and presented in a concise manner by Chadwick (1976).

The necessity to consider objective rates will then be clear, a preliminary step before introducing the incremental plasticity theory for finite transformations. The incrementally objective update algorithm, generalising the implicit algorithm presented in section 6.2, is finally presented.

Introducing the Concept of Objective Rates The kinematics of a typical finite transformation is illustrated in Figure 15. The studied domain occupies initially the volume Ω_0, at a time set arbitrarily to zero, and is convected to the volume Ω_t at time t. These domains initially and at time t are part of the reference (κ_0) and the current (κ_t) configurations, respectively. This transformation is described by the mapping $\underline{\phi}$ which is assumed to be invertible. The material point located initially at $\underline{X}$ in the reference configuration is convected to the point identified by the vector $\underline{x} = \underline{\phi}(\underline{X}, t)$ in the current configuration.

Consider a material fibre[4] $d\underline{X}$ of infinitesimal length in the reference configuration. This fibre is convected to $d\underline{x}$ by the transformation and the two vectors are related by the gradient of the transformation

$$d\underline{x} = \underline{\underline{F}}(\underline{X}, t) \cdot d\underline{X} \,. \tag{113}$$

This second-order tensor $\underline{\underline{F}}$ is neither Eulerian nor Lagrangian since it maps Lagrangian vectors to vectors found in the current configuration. Its determinant is called the Jacobian of the transformation and denoted J. The material element of volume dV_0 in the reference configuration is convected to the infinitesimal volume dV_t in the current configuration with the relation $JdV_0 = dV_t$. The initial Jacobian $J(\underline{X}, 0)$ takes thus the value of 1 since there is initially no volume change. Furthermore, at any time t, $J(\underline{X}, t)$ must be different from zero since the volume element dV_t cannot shrink to zero. This physical interpretation that J must always be positive, is equivalent to the statement that $\underline{\phi}$ is invertible at any time t.

This transformation gradient is now used to construct two important Lagrangian tensors, the dilation tensor and the Lagrangian deformation tensor. Any unit material vector $\underline{V}$ in the reference configuration is convected to the current configuration to the vector $\underline{v} = \underline{\underline{F}} \cdot \underline{V}$ such that its magnitude becomes $|\underline{v}| = (\underline{v} \cdot \underline{v})^{1/2} = [(\underline{V} \cdot {}^t\underline{\underline{F}}) \cdot (\underline{\underline{F}} \cdot \underline{V})]^{1/2}$. The square of this length is thus computed directly from the dilation tensor $\underline{\underline{C}}$ defined by

$$\underline{\underline{C}} = {}^t\underline{\underline{F}} \cdot \underline{\underline{F}} \,. \tag{114}$$

[4] A material fibre or vector should be seen as a marker which can be engraved on the material surface of or within the domain of interest.

It is a Lagrangian tensor since it acts on vectors defined in the reference configuration. The Lagrangian tensor of deformation is related to $\underline{\underline{C}}$ by

$$\underline{\underline{E}} = \frac{1}{2}(\underline{\underline{C}} - \underline{\underline{\delta}}), \tag{115}$$

and the scalar $2\underline{V} \cdot \underline{\underline{E}} \cdot \underline{V}$ is the square of the length of the convected vector $\underline{v}$ minus one.

Let's now introduce the velocity vector and its gradient. The velocity vector $\underline{U}$ is the Eulerian vector (defined in the current configuration) depicted in Figure 15 and defined as the time derivative of the mapping

$$\underline{U}(\underline{X}, t) = \frac{d}{dt}\underline{\phi}(\underline{X}, t). \tag{116}$$

This Eulerian vector is given a Lagrangian description since it is defined for all positions $\underline{X}$ in the reference configuration. It is often more convenient, especially in fluid mechanics, to describe the velocity field in terms of the current position vector $\underline{x}$. The velocity vector is then given by $\underline{U}(\underline{\phi}^{-1}(\underline{x}, t), t)$ using the inverse of the mapping $\underline{\phi}$. It is denoted simply $\underline{U}(\underline{x}, t)$, in this Eulerian description, without any risk of confusion between the two vectors having different descriptions. These two descriptions are now used to define the gradients of the velocity field in the reference and current configurations

$$d\underline{\dot{x}} = \text{Grad}(\underline{U}) \cdot d\underline{X} = \text{grad}(\underline{U}) \cdot d\underline{x}. \tag{117}$$

The Eulerian gradient grad($\underline{U}$) is also denoted $\underline{\underline{L}}$ and the Lagrangian gradient Grad($\underline{U}$) is nothing but $\dot{\underline{\underline{F}}}$ since the order of differentiation in time and space can be interchanged. The two tensors are related by

$$\underline{\underline{L}} = \dot{\underline{\underline{F}}} \cdot \underline{\underline{F}}^{-1}. \tag{118}$$

The tensor $\underline{\underline{L}}$ has the symmetric and the skew-symmetric part

$$\underline{\underline{D}} = \frac{1}{2}(\underline{\underline{L}} + {}^t\underline{\underline{L}}) \quad , \quad \underline{\underline{W}} = \frac{1}{2}(\underline{\underline{L}} - {}^t\underline{\underline{L}}), \tag{119}$$

which are the rate of deformation and the spin tensor, respectively. The former, $\underline{\underline{D}}$, is related to the time rate of change of the Lagrangian deformation tensor $\dot{\underline{\underline{E}}}$ by

$$ {}^t\underline{\underline{F}} \cdot \underline{\underline{D}} \cdot \underline{\underline{F}} = \dot{\underline{\underline{E}}}. \tag{120}$$

This relation between an Eulerian and a Lagrangian measure of the rate of deformation is important to define the various stress tensors needed in what follows.

The first stress tensor is Cauchy stress tensor $\underline{\underline{\sigma}}$ defined in the current configuration and which is central to the introduction of the stress vector $\underline{T}$ acting on any surface of normal $\underline{n}$ with $\underline{T} = \underline{\underline{\sigma}} \cdot \underline{n}$. This stress vector times the area dS of the facette of normal $\underline{n}$ is the force acting on that infinitesimal element of surface. The second stress tensor is Lagrangian and is called the second Piola-Kirchhoff stress tensor $\underline{\underline{S}}$. Its relation to Cauchy stress tensor is found by inspection of the equality of the mechanical power over an infinitesimal volume element in the reference and the current configuration

$$\underline{\underline{\sigma}} : \underline{\underline{D}}\, dV_t = \underline{\underline{S}} : \underline{\underline{\dot{E}}}\, dV_0 \,. \tag{121}$$

To derive this relation, consider first the relation (120) and recall that $dV_t = J dV_0$ and obtain

$$\underline{\underline{\tau}} = \underline{\underline{F}} \cdot \underline{\underline{S}} \cdot {}^t\underline{\underline{F}} \quad \text{with} \quad \underline{\underline{\tau}} = J \underline{\underline{\sigma}}\,, \tag{122}$$

in terms of yet another Eulerian stress tensor, $\underline{\underline{\tau}}$, the Kirchhoff stress tensor. It is this stress tensor which is used in the formulation of constitutive relations in a rate form. We shall need another stress tensor, the first Piola-Kirchhoff $\underline{\underline{P}}$ which transpose is ${}^t\underline{\underline{P}} = \underline{\underline{F}}^{-1} \cdot \underline{\underline{\tau}}$ in terms of the Kirchhoff stress and the transformation gradient. This stress tensor is neither Lagrangian nor Eulerian, as the transformation gradient. In fact, it is conjugate to the time rate of change of the transformation gradient in the sense of power used in equation (121), since

$$\underline{\underline{D}} : \underline{\underline{\sigma}}\, dV_t = \underline{\underline{L}} : \underline{\underline{\sigma}}\, dV_t = \underline{\underline{\dot{F}}} \cdot \underline{\underline{F}}^{-1} : \underline{\underline{\sigma}} J\, dV_0 = \underline{\underline{\dot{F}}} : (\underline{\underline{F}}^{-1} \cdot \underline{\underline{\tau}})\, dV_0 \,. \tag{123}$$

Having defined the various stress tensors, we turn our attention to the construction of an objective Eulerian stress rate. We do concentrate on stresses although the argument to be developed is applicable to any Eulerian tensor. The differentiation of the two sides of (122) provides

$$\underline{\underline{\dot{\tau}}} = \underline{\underline{L}} \cdot \underline{\underline{\tau}} + \underline{\underline{\tau}} \cdot {}^t\underline{\underline{L}} + \underline{\underline{F}} \cdot \underline{\underline{\dot{S}}} \cdot {}^t\underline{\underline{F}}\,, \tag{124}$$

between the rates of change of the second Piola-Kirchhoff stress and of the Kirchhoff stress. This relation is crucial to explain the definition and our choice of objective stress rate. Constitutive relations are formulated such that the material response is invariant by a change of frame of the observer. A complete discussion is beyond the scope of this contribution and the interested reader is referred to the classical textbook of Malvern (1969) for a well illustrated presentation. Two observers, the first in the reference configuration and the second in the current configuration should estimate the same stress changes. If the first observer does not notice any stress change in the reference configuration $\dot{\underline{\underline{S}}} = \underline{\underline{0}}$, then the appropriate stress rate in the current configuration of the Kirchhoff stress should also be zero and thus cannot be given by (124). The following stress rate of the Kirchhoff stress is objective

$$\overset{T}{\underline{\underline{\tau}}} \equiv \underline{\underline{F}} \cdot \dot{\underline{\underline{S}}} \cdot {}^t\underline{\underline{F}} = \dot{\underline{\underline{\tau}}} - \underline{\underline{L}} \cdot \underline{\underline{\tau}} - \underline{\underline{\tau}} \cdot {}^t\underline{\underline{L}}, \tag{125}$$

and is appropriate for describing constitutive relations. It is referred to in what follows as Truesdell's rate and is marked by the superposed letter T. Note that other objective rates could be proposed and we limit ourselves to the rate which will be used in the numerical scheme. Note also that the result obtained above means that any rate of a Lagrangian tensor, say $\dot{\underline{\underline{A}}}$, is convected to an objective rate in the current configuration defined by the relation $\underline{\underline{F}} \cdot \dot{\underline{\underline{A}}} \cdot {}^t\underline{\underline{F}}$.

The rest of this section is now devoted to the theorem of virtual work, which extends (107) to the context of finite transformations, and its linearisation for the finite-element equilibrium search. The virtual internal work is defined by ${}^t\underline{\underline{P}} : \hat{\underline{\underline{F}}}$ per unit of volume in the reference configuration and the total internal virtual work is set equal to the external virtual work

$$\int_{\Omega_0} {}^t\underline{\underline{P}} : \hat{\underline{\underline{F}}}\, dV_0 \quad = \quad \int_{\Omega_0} \rho_0 \underline{g} \cdot \hat{\underline{u}}\, dV_0 + \int_{\partial\Omega_0} \underline{T}^d \cdot \hat{\underline{u}}\, dS\,, \tag{126}$$

in which ρ_0 is the volumetric mass in the reference configuration. The Kirchhoff stress, central to the constitutive relations, is introduced in the internal virtual work in the left-hand side of (126) which becomes

$$\int_{\Omega_0} \underline{\underline{\tau}} : \hat{\underline{\underline{\eta}}}\, dV_0 \quad \text{with} \quad \hat{\underline{\underline{\eta}}} = \frac{1}{2}(\hat{\underline{\underline{F}}} \cdot \underline{\underline{F}}^{-1} + {}^t\underline{\underline{F}}^{-1} \cdot {}^t\hat{\underline{\underline{F}}})\,, \tag{127}$$

in which $\hat{\underline{\underline{\eta}}}$ is the virtual Eulerian strain tensor having the same structure as the rate of deformation tensor in (119) if the time operator is replaced by the superposed hat which marks virtual quantities. The weak form of equilibrium proposed by the theorem of virtual work is at the basis of the finite-element method for non-linear problems. The non-linearities are now due to material properties and to the geometry. The solution is in terms of the displacement field once the finite-element discretisation has been introduced. This discretisation step, well understood for the linear problems, is avoided here for sake of clarity. The search with Newton-Raphson for the equilibrium solution requires a linearisation of the theorem of virtual work. This linearisation reads

$$\int_{\Omega_0} \left(\underline{\hat{\eta}} : \frac{\partial \underline{\underline{\tau}}}{\partial \underline{\underline{E}}} : \underline{\underline{\tilde{E}}} + \underline{\underline{\tau}} : \underline{\underline{\tilde{\hat{\eta}}}} \right) dV_0 = \tag{128}$$

$$\int_{\Omega_0} \rho_0 \underline{g} \cdot \underline{\hat{u}} \, dV_0 + \int_{\partial\Omega_0} \underline{T}^d \cdot \underline{\hat{u}} - \int_{\Omega_0} \underline{\underline{\tau}} : \underline{\underline{\hat{\eta}}} \, dV_0 \,,$$

in terms of the increment in Lagrangian deformation $\underline{\underline{\tilde{E}}}$. Note the extra term in the left-hand side due to the non linearity of the virtual Eulerian strain tensor. These geometrical terms $\underline{\underline{\tilde{\hat{\eta}}}}$ are expressed in terms of the Eulerian gradient of the virtual $\underline{\underline{\nabla \hat{u}}}$ and the incremental $\underline{\underline{\nabla \tilde{u}}}$ displacement field such that

$$\underline{\underline{\tau}} : \underline{\underline{\tilde{\hat{\eta}}}} = -\underline{\underline{\tau}} : \frac{1}{2} (\underline{\underline{\nabla \tilde{u}}} \cdot \underline{\underline{\nabla \hat{u}}} + {}^t\underline{\underline{\nabla \hat{u}}} \cdot {}^t\underline{\underline{\nabla \tilde{u}}}) \,. \tag{129}$$

Plasticity Theory Having selected the Kirchhoff stress to define constitutive relations, we postulate that the elastic domain can be described in terms of this stress tensor

$$E_\tau = \{ \underline{\underline{\tau}} \in \mathbb{R}^6 \quad \text{with} \quad f(\underline{\underline{\tau}}, \gamma^p) \leq 0 \} \,, \tag{130}$$

selecting one more time the equivalent plastic strain γ^p as the unique variable to describe the evolution of the elastic domain. Consider now the additive decomposition of the rate of deformation tensor

$$\underline{\underline{D}} = \underline{\underline{D}}^e + \underline{\underline{D}}^p \,, \tag{131}$$

in an elastic and a plastic part, which generalises to finite transformation the decomposition made previously for multi-axial plasticity of the linearised deformation tensor. Note that there are numerous plasticity theories for finite deformation which would favour instead of (131) a multiplicative decomposition of the transformation gradient ($\underline{\underline{F}} = \underline{\underline{F}}^e \cdot \underline{\underline{F}}^p$) which we will not use here. The multiplicative decomposition is strongly motivated by the microstructural properties of crystals but is more difficult to justify for frictional and cohesive materials. It is for this reason that the additive decomposition of the rate of deformation tensor (131) is favoured here. This choice leads to an elastic rate of deformation related linearly to Truesdell's rate of the Kirchhoff stress by

$$\overset{T}{\underline{\underline{\tau}}} = \mathbb{D}^e : (\underline{\underline{D}} - \underline{\underline{D}}^p) \,, \tag{132}$$

which is referred to as an hypo-elasticity theory. The material properties entering the operator $\mathbb{D}^e$ are defined in the reference configuration and not in the current configuration as it should be done rigorously. The error

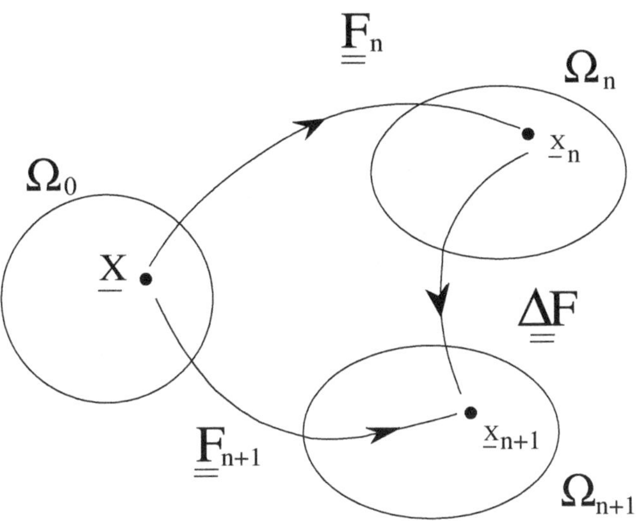

Figure 16. The studied domain occupies the volume Ω_o in the reference configuration and Ω_t and Ω_{t+1} in the two current configurations separated by the time increment Δt. The transformation gradient $\underline{\underline{\Delta F}}$ links these two successive configurations and is obtained by the combination of the inverse of the transformation from Ω_0 to Ω_n with the transformation from Ω_0 to Ω_{n+1} resulting in $\underline{\underline{\Delta F}} = \underline{\underline{F}}_{n+1} \cdot \underline{\underline{F}}_n^{-1}$.

could be estimated during a cycle of elastic straining at finite strain where an hysteresis in the stress is detected. The error is however marginal if the amount of elastic deformation is small compared to the plastic deformation, an hypothesis met in most of the applications of interest here.

The last point required to complete this short presentation of the plasticity theory for finite transformations is the flow rule. In view of the structure adopted for describing the elastic domain, it seems reasonable to describe the flow potential in terms of the same stress measure and the flow rule then reads

$$\underline{\underline{D}}^p = \dot{\gamma}^p \frac{\partial g}{\partial \underline{\underline{\tau}}}, \tag{133}$$

in terms of the flow potential g and the Kirchhoff stress.

Update Algorithms The objective of this section is to present an update algorithm based on incremental quantities constructed from objective strain and stress incremental tensors. The basic idea learned from the previous discussion on objective rates is that a Lagrangian tensor rate, say $\dot{\underline{\underline{A}}}$, convected to the current configuration as $\underline{\underline{F}} \cdot \dot{\underline{\underline{A}}} \cdot {}^t\underline{\underline{F}}$, is objective.

Time is now discretised and the current configurations at two successive times t_n and t_{n+1} are presented in Figure 16. The material point initially at $\underline{X}$ is found at $\underline{x}_n$ and $\underline{x}_{n+1}$ at these two times. The incremental transformation gradient $\underline{\underline{\Delta F}}$ characterises the transformation between these two points. It is constructed by, first, considering $\underline{\underline{F}}_n^{-1}$ to link the domain Ω_n to the domain Ω_0 in the reference configuration and, second, applying the gradient $\underline{\underline{F}}_{n+1}$ to Ω_{n+1} in the current configuration. This combination reads

$$\underline{\underline{\Delta F}} = \underline{\underline{F}}_{n+1} \cdot \underline{\underline{F}}_n^{-1} . \tag{134}$$

We now define an incremental Lagrangian measure of deformation by proposing to express the rate of dilation, defined in (114), as the difference

$$\dot{\underline{\underline{C}}}_{n+1} = \frac{1}{\Delta t}\left(\underline{\underline{C}}_{n+1} - \underline{\underline{C}}_n\right) . \tag{135}$$

Introduce the definition (134) and express this rate as

$$\dot{\underline{\underline{C}}}_{n+1} = \frac{1}{\Delta t}\,{}^t\underline{\underline{F}}_n \cdot \left(\underline{\underline{\Delta C}} - \underline{\underline{\delta}}\right) \cdot \underline{\underline{F}}_n \quad \text{with} \quad \underline{\underline{\Delta C}} = {}^t\underline{\underline{\Delta F}} \cdot \underline{\underline{\Delta F}} . \tag{136}$$

The rate of deformation tensor $\underline{\underline{D}}_{n+1}$ is then deduced from this relation once the definitions (115) and (120) are accounted for:

$$\underline{\underline{D}}_{n+1} = \frac{1}{2\Delta t}\,{}^t\underline{\underline{\Delta F}} \cdot \left(\underline{\underline{\Delta C}} - \underline{\underline{\delta}}\right) \cdot \underline{\underline{\Delta F}} . \tag{137}$$

Let's check the objectivity of this rate of deformation tensor by considering a finite rigid body motion between the configurations at time t_n and t_{n+1} characterised by the rotation $\underline{\underline{Q}}$ (recall the property ${}^t\underline{\underline{Q}} \cdot \underline{\underline{Q}} = \underline{\underline{\delta}}$). The incremental tensor $\underline{\underline{\Delta C}}$ is then $\underline{\underline{\delta}}$ according to the definition in (136) and the Eulerian rate of deformation in (137) is indeed equal to zero. The incremental construction of the kinematics is thus objective.

The increment in stress is also defined in the reference configuration. The second Piola-Kirchoff stress rate is

$$\dot{\underline{\underline{S}}}_{n+1} = \frac{1}{\Delta t}\left(\underline{\underline{S}}_{n+1} - \underline{\underline{S}}_n\right) . \tag{138}$$

Its relation to the Kirchhoff stress at the times t_n and t_{n+1} is then

$$\dot{\underline{\underline{S}}}_{n+1} = \frac{1}{\Delta t}\underline{\underline{F}}_n^{-1} \cdot \left(\underline{\underline{\Delta F}}^{-1} \cdot \underline{\underline{\tau}}_{n+1} \cdot {}^t\underline{\underline{\Delta F}}^{-1} - \underline{\underline{\tau}}_n\right) \cdot {}^t\underline{\underline{F}}_n^{-1}\,, \tag{139}$$

making use of the relation in (122) and the definition (134). The Truesdell's rate of the Kirchhoff stress introduced in (125) is defined at time t_{n+1} by

$$\overset{T}{\underline{\underline{\tau}}}_{n+1} = \underline{\underline{F}}_{n+1} \cdot \dot{\underline{\underline{S}}}_{n+1} \cdot {}^t\underline{\underline{F}}_{n+1}\,, \tag{140}$$

or equivalently by

$$\overset{T}{\underline{\underline{\tau}}}_{n+1} = \frac{1}{\Delta t}\left(\underline{\underline{\tau}}_{n+1} - \underline{\underline{\Delta F}} \cdot \underline{\underline{\tau}}_n \cdot {}^t\underline{\underline{\Delta F}}\right), \tag{141}$$

using (138) and (134).

If the transformation from Ω_n to Ω_{n+1} is a rigid body motion characterised by the rotation $\underline{\underline{Q}}$ what is this Truesdell's rate ? In that instance, the stress $\underline{\underline{\tau}}_n$ could be seen as the stress in the reference configuration at time t_n and thus interpreted as the second Piola-Kirchhoff stress for an observer attached to the configuration at time t_{n+1}. The relation between $\underline{\underline{\tau}}_n$ and $\underline{\underline{\tau}}_{n+1}$ in the configuration at time t_{n+1} is thus analogous to (122) and reads $\underline{\underline{\tau}}_{n+1} = \underline{\underline{\Delta F}} \cdot \underline{\underline{\tau}}_n \cdot {}^t\underline{\underline{\Delta F}}$ since the transformation gradient between the two successive configurations is $\underline{\underline{\Delta F}}$. Consequently, the Truesdell's rate of the Kirchhoff stress in (140) is indeed equal to zero during an incremental rigid body motion. Truesdell's stress rate for an increment Δt is thus objective.

The rest of this section is devoted to the update algorithm based on these two objective rates. The generalisation of (100) reads

$$\begin{aligned} \overset{T}{\underline{\underline{\tau}}}_{n+1} &= \mathbb{D}^e : (\underline{\underline{D}}_{n+1} - \Delta\gamma^p_{n+1}\underline{\underline{M}}_{n+1})\,, \\ \overset{T}{\underline{\underline{\tau}}}_{n+1} &= \frac{1}{\Delta t}\left(\underline{\underline{\tau}}_{n+1} - \underline{\underline{\Delta F}} \cdot \underline{\underline{\tau}}_{n+1} \cdot {}^t\underline{\underline{\Delta F}}\right), \\ & f(\underline{\underline{\tau}}_{n+1}, \gamma^p_{n+1}) \leq 0\,, \end{aligned} \tag{142}$$

where the flow rule, estimated at time t_{n+1} is already used to define the plastic part of the rate of deformation tensor. This system of equations has for unknowns the accumulated plastic strain γ^p_{n+1} and the Kirchhoff stress $\underline{\underline{\tau}}_{n+1}$ at the end of the increment. The solution is found with the two-steps procedure outlined above. The elastic predictor step assumes that no plastic flow occurs during the increment $\Delta\gamma^p_* = 0$ and the resulting, predicted stress is

$$\underline{\underline{\tau}}_* = \underline{\underline{\Delta F}} \cdot \underline{\underline{\tau}}_{n+1} \cdot {}^t\underline{\underline{\Delta F}} + \mathbb{D}^e : \underline{\underline{\Delta e}} \quad \text{with} \quad \underline{\underline{\Delta e}} = \frac{1}{2}\left(\underline{\underline{\delta}} - {}^t\underline{\underline{\Delta F}}^{-1} \cdot \underline{\underline{\Delta F}}^{-1}\right).$$

This predictor stress, or its invariants are used to check the criterion (142c). If it is satisfied, no plasticity indeed occurred and the predictor stress is the final stress state. If the opposite conclusion is reached, the correction step should be applied to obtain the final stress

$$\underline{\underline{\tau}}_{n+1} = \underline{\underline{\tau}}_{*} - \Delta\gamma^{p}_{n+1}\, \mathbb{D}^{e} : \underline{\underline{M}}_{n+1} \,. \tag{143}$$

The procedure for this correction step is identical to the one followed for the small-strain formulation. The criterion is enforced at the end of the time step and all stress variables are expressed in terms of the increment in equivalent plastic strain $\Delta\gamma^{p}_{n+1}$. This unknown is determined using an iterative method and the rest of the update is conducted with the classical radial return for the deviatory stress.

The last item which should be discussed is the linearisation of this update algorithm to derive the consistent tangent. This step is rather technical because the resulting expressions are not a mere generalization of (128) where the strain measure would now be the incremental Lagrangian strain. It turns out that the linearised theorem of virtual work is expressed in terms of the symmetric part and the skew-symmetric part of the Eulerian gradient of the incremental displacement. The symmetric part was expected but the skew-symmetric part, the incremental spin, adds a certain complexity to the writing. Its account does not require any new concept but the derivation of the final expression for the consistent tangent becomes a burden which will not be shared with the reader ! It is preferred to conclude this section with some applications which show the merits of the so-called spectral elements.

6.4 Applications

The two 2D applications assume plane-strain conditions and make use of spectral elements. The reader is invited to solve the problem of section 7.3 to discover this class of elements. It suffices to know that quadrature points (Lobatto's rule) and the nodes share the same position. The shape functions are Lagrange polynomials. The two problems consist of the compression of a rectangular-shaped specimen at the laboratory scale, and the buckling of a competent plate at the kilometre-scale. The competent material in the two examples is frictional and cohesive, its plasticity being described by the Drucker-Prager criterion. The plastic deformation is isochoric, there is thus no irreversible volume change and the flow potential is thus the classical Von Mises equivalent stress. The update algorithm does correspond to the one discussed in the previous section. A viscosity is introduced, more for numerical stability than for its physical relevance, in the form of the following power-law function

$$\dot{\gamma}^p = \dot{\gamma}_0\Big(\Big[\frac{\sigma^e_{DP}}{c(\gamma^p)}\Big]^m - 1\Big)\,, \tag{144}$$

in which $\dot{\gamma}_0$ and m are the reference strain rate and the strain-rate exponent, respectively. The two stresses σ^e_{DP} and $c(\gamma^p)$ in (144) are the Drucker-Prager equivalent stress defined in (90) (in terms of the Kirchhoff stress and with a different scaling of J_2 which is done for a uni-axial stress state in the present implementation) and the cohesion, respectively. The work-hardening is introduced via the cohesion which is a function of the accumulated plastic strain, according to either one of the two functions

$$\begin{aligned}
(1)\quad c(\gamma^p) &= C_0\Big(1 + (\frac{\gamma^p}{\gamma_0})^n\Big)\,; && (145)\\
(2)\quad c(\gamma^p) &= C_I + (C_M - C_I)2\frac{\sqrt{\gamma_M\gamma^p}}{\gamma^p + \gamma_M}, \quad \text{for} \quad \gamma^p \leq \gamma^*\,, \\
c(\gamma^p) &= C_F + (c(\gamma^*) - C_F)\exp\Big(\frac{c'(\gamma^*)}{c(\gamma^*) - C_F}(\gamma^p - \gamma*)\Big)\,, \\
&\quad \text{for} \quad \gamma^p > \gamma^*\,.
\end{aligned}$$

The first function is the power law, a monotonically increasing function of the accumulated plastic strain. The second function involves hardening and softening and is C_1 despite the introduction of two expressions over the intervals $[0, \gamma^*]$ and $]\gamma^*, +\infty]$. The function increases from the initial cohesion C_I to reach the maximum C_M for the accumulated plastic strain γ_M. The function then decreases. To ensure a smooth evolution towards the residual cohesion C_F at large strain, the second expression is proposed in (145) for γ^p greater than the transition strain γ^*. The various material parameters and their values selected for the two examples are summarised in Table 5.

Strain Localization The first problem consists of the plane-strain compression of a rectangular specimen. The specimen width to length ratio is 1 to 2 and the characteristic size of the order of centimetres, a length typical of the laboratory specimen. Gravity effects are thus disregarded and the initial stresses are assumed to be zero. Compression is achieved by a stiff plateau such that the vertical displacement of all points on the top section of the specimen are identical and controlled by the loading device. The rate of displacement of the upper plateau leads to a nominal strain rate of $5 \times 10^{-4} s^{-1}$. There is no friction between the plateau and the specimen. The specimen is placed on a rigid base, fixed in space, with again a zero

Parameter	Definition	Values & unit
	Elastic and frictional properties	
E	Young's modulus	10 GPa
ν	Poisson's ratio	0.2
μ	friction coefficient	0.6
	Power law hardening (eq 145a)	
C_0	initial yield stress	5 MPa
γ_0	reference strain	1 %
n	hardening exponent	0.8
	Hardening & softening function (eq 145 b and c)	
C_I	initial cohesion	1 MPa
C_M	maximum cohesion	5 MPa
γ_M	equivalent strain at σ_M	10 %
γ^*	transition strain	50 %
C_F	final cohesion	0.1 MPa
	Viscosity: power law	
$\dot{\gamma}_0$	reference strain rate	1 $\%/s$
m	strain-rate exponent	10
	Geometry of the buckling problem	
D	plate depth	1 km
w	plate width	20 m
L	plate length	200 m
	Further material properties	
ρ_s	material density for competent layer	2300 kg/m^3
ρ_f	material density for surrounding sediments	2300 kg/m^3

Table 5. Parameters for the elasto-plastic rock considered for the two examples and the geometry of the initial buckling plate.

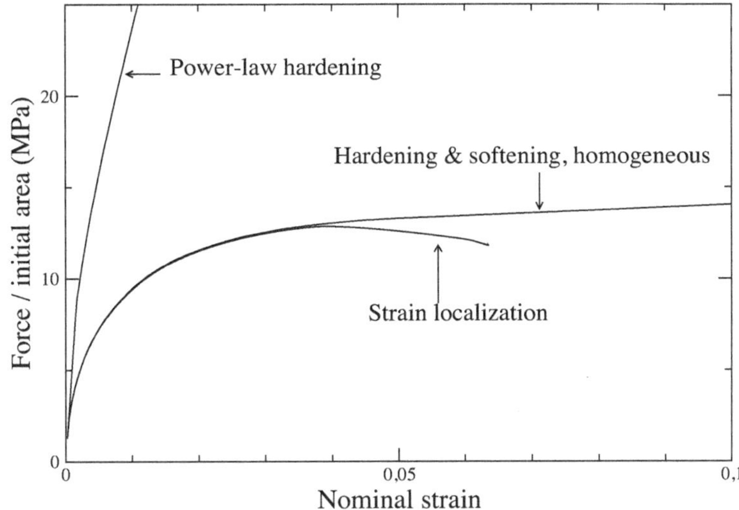

Figure 17. The load-displacement relation for the plane-strain compression of a rectangular block.

friction at the boundary. An imperfection is introduced in the geometry of the specimen by reducing linearly its width from the top to the bottom by 2%, the left vertical side remaining perpendicular to the two plateaux. A *single* spectral element composed of 8×8 nodes is first employed for the spatial discretisation.

The first numerical test is based on the power-law hardening function (145a) and the response in terms of load as a function of the nominal strain (the displacement of the top plateau divided by the specimen length) is presented in Figure 17. There is a continuous increase of the force, with or without the geometrical imperfection, and the deformation remains homogeneous during the whole test. No geometrical or material instabilities were detected even at 30 % of nominal strain. The second and third numerical tests are performed with the hardening and softening function (145b and c). The force nominal strain curve in the absence of any imperfection (the deformation is homogeneous) is presented also in Figure 17. The increase in specimen width during the compression leads to a force increase which counterbalances the softening which is introduced by the constitutive response. However, the important decrease in the tangent of this curve points to potential instabilities in the structure response. The concept of instabilities is used here in a loose sense meaning a change in the mode of deformation from the homogeneous (fundamental) solution obtained in the absence of imperfection. It is hoped that the proposed imperfection is appropriate to capture the development of these various instabilities. Several geometrical instabilities occur close to the fundamental solution of this problem and they, characteristically, break the symmetry of the solution in displacement. The first mode of geometrical instability is likely to have a barrel shape and the second a S shape. They break the symmetry with respect to the horizontal axis and with respect to both the horizontal and the vertical axes passing through the specimen geometrical centre, respectively. However, the severe reduction in the tangent modulus leads to the onset of yet another mode of instability which is not purely geometrical but the result of the material response. It is the shear band instability, which becomes dominant and which is responsible for the final failure of the specimen by the localised deformation in a narrow band. These various modes develop because of the slight perturbation in the otherwise homogeneous solution due to the geometrical imperfection. This development becomes important when the load nominal strain curve deviates from the homogeneous solution, Figure 17.

The calculations were stopped at the stage of development of the shear band instability corresponding to the distribution of accumulated plastic strain γ^p shown in Figure 18a. The level of strain in dark blue, 5.3% gives us an idea of the loading stage when the deformation ceased to be homogeneous and localization is initiated. The two block around the shear band behave more or less rigidly, with no further accumulation of strain. The deformation is indeed localised within a narrow band, delineated approximately by the black isocontours. The band width is set by the distance between two to three nodes within the element, shown with solid dots. This distance is thus set by the mesh size. Plastic flow is active within the band while an elastic unloading occurred during the strain localization outside the band. This unloading initiated away from the actual band centre. Note that the band direction is expected at $\pi/4 + \phi/2$ from the horizontal direction for the classical Coulomb criterion with $\mu = \tan\phi$ where ϕ is the friction angle. This angle is 31° in this calculation and thus the band theoretical orientation is 60°, approximately. The observed orientation is closer to 50° and seems to be controlled by the yet plastically deforming nodes within the element. This is certainly the best approximation which one could get with a single element but this remark calls on a discussion on the ability of finite elements

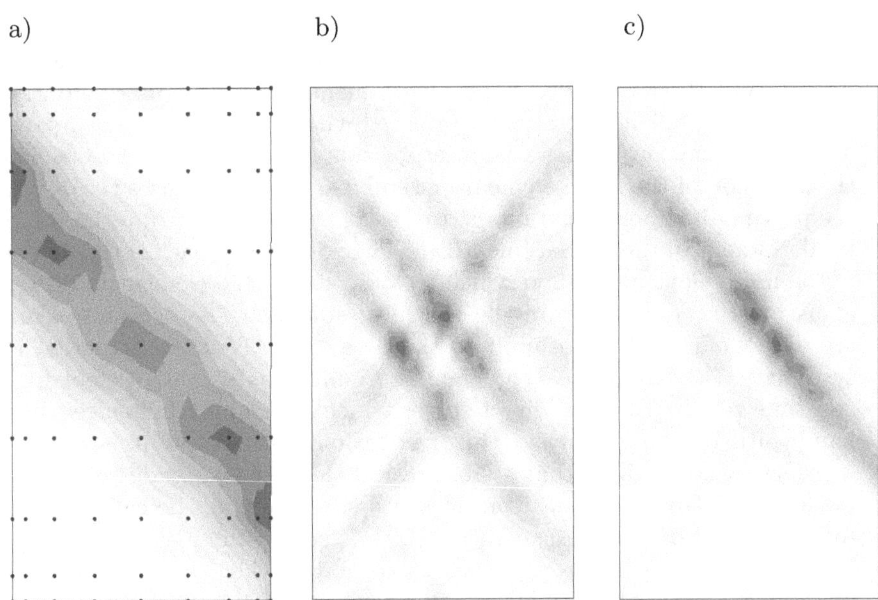

Figure 18. Ten white to black isocontours of accumulated plastic strain due to localization during the compression test. In a), a single spectral element 8×8 (the solid dots are the nodes) is considered and the strain range is [5.3%; 25.5%]. In b) and c), the mesh is composed of 4×4 elements of 8×8 nodes. The strain ranges are [2.%; 14.2%] and [2.%; 35.2%] in b) and c).

in capturing strain localization.

Finite-element with few nodes such as the four-noded Lagrange elements, which were popular in the eighties, are not meant to capture sharp gradients in the strain. This difficulty was recognised by Tvergaard et al. (1981) who proposed the use of four-crossed triangles, which we considered in section 2, with the slight variation that the mean stress was averaged over each patch of four triangles, for reasons which we will not discuss here. Their mesh was designed such that the diagonals of the deformed patches were coinciding with the direction of shear banding at the appropriate level of deformation. These diagonal directions are the most prompt to develop discontinuities, as it was discussed in section 2 and shown in Figure 9b and c with the two stress distributions. Classical elements of the Lagrange family and especially the four-noded element do no have this interesting property. It is for that reason that Ortiz et al. (1987) proposed to modify the shape functions of the four-noded Lagrange element to include discontinuities in the velocity gradient consistently oriented with the shear band at its onset. This enhancement of the spatial interpolation should be seen as a member of the p version of the finite-element method where one increases the degree of the polynomial approximation, p, over each element to capture the complex local gradient in the displacement field. The other approach is the h version of the finite-element method which sees the size of the element to decrease, at a constant p, to capture the spatial gradients. The spectral elements are without any doubt prompt to the p version. Their construction is so systematic that the increase in the node number is painless except for the cpu-time required by the direct solver which is strongly influenced by the bandwidth of the global system of equations. However, it is extremely satisfactory to see here that a single 8×8 element is already so successful in capturing the strain localization process. Of course, the calculation has to be stopped once the shear band width is set by the mesh and the band orientation is constrained by the set of nodes which are still plastically deforming. These two problems, the shear band width set by the mesh and oriented by the deforming nodes, are nevertheless always present in any finite-element analysis of strain localization.

The strain localization analysis is finalised by considering a finer mesh of 4×4 elements of 8×8 nodes. There is thus a total of 841 nodes. The results of the simulation are presented in Figure 18b and c, corresponding to two stages of the deformation where the maximum accumulated plastic deformation is 14.2% and 35.2%, respectively. In the first graph, the deformation has already ceased to be homogeneous and several conjugate shear bands are in competition. The level of straining in the regions behaving more or less rigidly (light grey) is only 2% showing that the localization onset is much earlier than expected from the results of the single element analysis. As loading continues, the structure selects one of the shear band to develop. This shear band defines the final failure mode of the specimen. This band is rather close to the one observed in Figure 18a for the coarse mesh in terms of orientation and position in the specimen. It is much thinner because of the finer mesh and its thickness is still set by the distance between two or three neighbouring nodes.

Folding of an Elasto-Plastic Beam The second problem concerns the buckling of an elasto-plastic beam. The beam is surrounded by soft sediments, typically shales, which are described as inviscid fluids of material density ρ_f. The beam of length L width w and material density ρ_s is at the

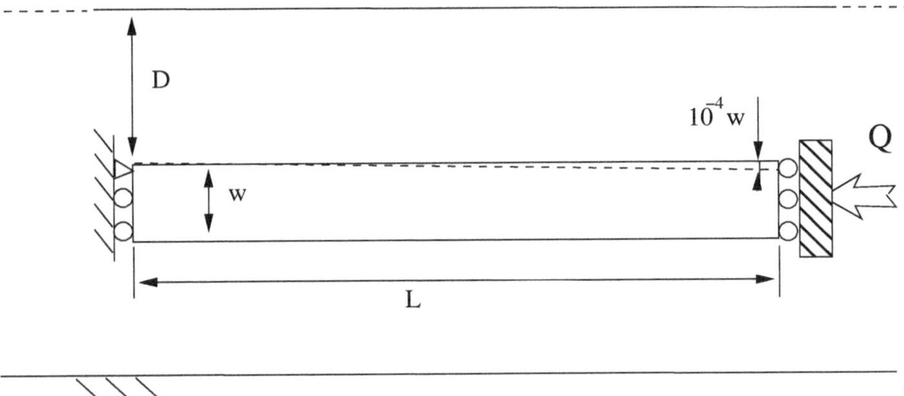

Figure 19. The beam in compression is surrounded by soft sediments described as inviscid fluids. The imperfection is defined by a reduction of the cross section to the right by 10^{-4} time the width w.

depth D, Figure 19. The nodes on the two lateral boundaries are free to move vertically except for the top left corner node which is fixed in space at the depth D. There are thus no shear forces along these two lateral contact surfaces. A force is applied on the right boundary to produce a displacement towards the left at a nominal strain rate of $2 \times 10^{-4} s^{-1}$. The fluid-like material acts on the solid beam by applying a compressive force distribution directed normal to the top and bottom surfaces. The force intensity is equal to the local lithostatic pressure. The total height of fluid D is kept constant in time. Note that a geometrical imperfection is introduced and corresponds to the linear variation in the cross section of the beam which is reduced by 10^{-4} from the left to the right, Figure 19. The domain is discretised with a single line of five elements. Each element is a 6×6 spectral element.

The first numerical test is conducted for an elastic plate. This material response is obtained by setting the cohesions to unrealistic large values. The material is thus hypo-elastic, a minor error in what follows. The load nominal strain is presented in Figure 20. The response is linear and the strain close to homogeneous prior to the Euler load marking the drastic change in slope of this curve. Beyond this load, the shortening of the beam is conducted with a minor increase in the applied load. The deformation of the beam ceases to be homogeneous and the plate buckles up in a mode which is identical to the one described next for elasto-plastic materials.

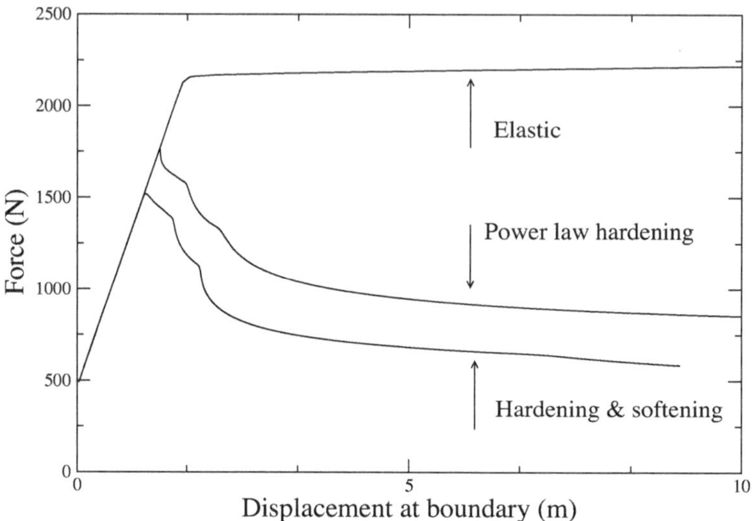

Figure 20. The load-displacement relation for the buckling of the elasto-plastic plate. The plane-strain compression test presented in Figure 19 illustrate the two definitions of the power-law hardening and of the hardening and softening law adopted for the cohesion.

The second run is conducted for the cohesive, frictional material described above assuming the power-law hardening function in (145a). The load-nominal strain response is presented in Figure 20. Plasticity occurs before the applied load reaches the Euler load. Plasticity is accompanied by a sufficient reduction in the tangent stiffness such that the buckling follows immediately first yield. The load beyond that buckling is found to decrease with increasing shortening. This is a consequence of the pressure-sensitivity of the Drucker-Prager equivalent stress. The shape of the buckled fold after the onset of buckling is presented in Figure 21a. The amplitude of the fold develops as the loading is pursued, Figure 21b and c. There is partial unloading in the central region of the beam marked on the load-displacement curve 20 by abrupt changes in the slope. The over-all shape of the beam remains close to a sine wave despite the unloading. The white to black isocontours of the accumulated plastic strain plotted on the deformed configuration reveals that the central region of the structure is responding mostly elastically and that the deformation is restricted to the first and the

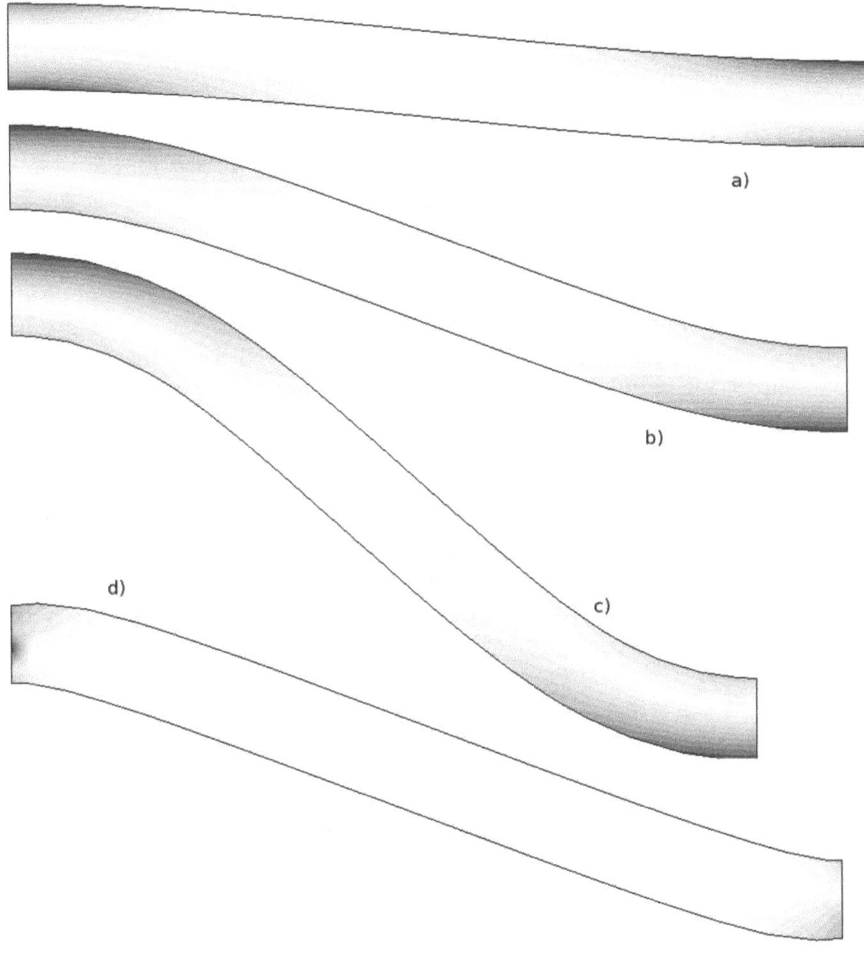

Figure 21. The buckled beam at three different stages of the compression, a) to c), assuming the power-law hardening presented in Figure 17. Figure d) corresponds to the buckled plate assuming the hardening & softening relation also presented in Figure 17. In all cases, the distribution of equivalent plastic strain is presented with white to black isocontours for values ranging from zero to 2.8, 16.6, 32.5 and 96.6 % for a) to d), respectively.

last quarter of the structure length.

The third and last run is conducted for the hardening and softening function (145b and c). There is thus, from the analysis of the compression test discussed previously, a tendency for the strain to localize. This trend is observed in Figure 21d where, despite the small amplitude of the fold, the majority of the deformation is found to be localised in two hinges, one at each end of the beam. The limb between the hinges is rather straight. The deformation is elastic in the limb (complete unloading) and the deformation is localised to a few nodes in the two hinges. The mesh is of course too coarse to resolve any band distribution in these two regions. In conclusion, it appears that the gentle fold obtained with the power-law rheology, has given place to a kink fold or chevron, in agreement with the interpretation of Massin et al. (1996), if the material response is conducive to strain localization.

7 Exercises and a Problem

It is impossible to learn a subject without practicing. This section provides such opportunity with two exercises and one problem. The first exercise is on the Patch test, discussed in section 5. The second exercise is the extension to visco-plasticity for 1D Von Mises materials of the update algorithm discussed in section 6. The problem concerns the construction of the spectral elements which are used in a 2D setting for the two applications of section 6. These exercises and the problem are based on a series of exams given at Ecole Polytechnique, Palaiseau, France, in connection with the course on the finite-element method offered by Bonnet and Frangi (2005). Solution hints are provided in the next section.

7.1 Exercise 1 : The Patch Test.

The patch test proposed to study the convergence of incompatible elements by Strang (1972) following the pioneering idea of Irons (Bazeley et al., 1965) is considered here as a validation step during the development of a finite-element code. The transformation is infinitesimal, the material linear, isotropic elastic and body forces are disregarded.

Consider first the example of the mesh in Figure 22a composed of four elements which are four-noded and member of the Lagrange family. The nodal displacements on the boundary (all nodes except node no 5) are imposed according to the relation

$$u_i = \alpha_i + \beta_{ij} x_j \, , \tag{146}$$

where x_j is the nodal coordinate and α_i and β_{ij} compose a set of six arbi-

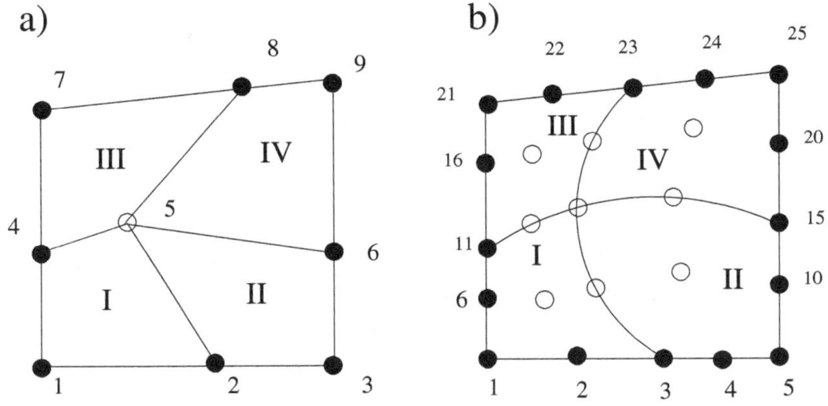

Figure 22. Two meshes composed of four elements which are, four-noded in a) and nine-noded in b) and members of the Lagrange family. The Patch Test is conducted by assigning nodal displacement at the nodes on the domain boundary (the solid dots). The nodal displacement within the domain (open circle) is determined by the finite-element approach.

trary constants.

1• What is the exact, analytical solution to this problem if (146) is consistently applied at the boundary ?

2• What is the connectivity table of the mesh in Figure 22a using the local nodal numbering shown in Figure 12 ?

3• What should the displacement at node 5 (global numbering), of coordinate $(x_1^{(5)}, x_2^{(5)})$ be ?

4• What should the linearised strain be at every quadrature point of the mesh ?

5 • The same type of boundary conditions (146) are now applied to the nodes on the boundary of the mesh in Figure 22b. Provide the connectivity table for this second mesh. What is the displacement at the central nodes ?

7.2 Exercise 2: Visco-Plasticity of a 1D Von Mises Material

Most engineering materials for rates of straining larger than $10^{-3}s^{-1}$ and ambient temperature present a plastic deformation which is function of the rate of loading. Natural materials deeper than the brittle-ductile transition of the Earth's crust are also strain-rate dependent. Experimental results were presented in Figure 13b for the traction of a bar. We keep with this 1D setting and propose to extend the algorithm discussed in section 6.1. The elasticity modulus is denoted E and the limit of elasticity described by the Von Mises criterion:

$$f(\sigma, \gamma^p) = \sigma^e - \sigma_Y(\gamma^p) \leq 0\,, \tag{147}$$

in which $\sigma^e = |\sigma|$ is the equivalent Von Mises stress and σ_Y the hardening function. The positive scalar γ^p is the accumulated plastic deformation. The hardening function could be constructed by conducting a test at sufficiently small nominal strain rate and would correspond to the curve CD of the tests presented in Figure 13b.

The plastic deformation if given by

$$\dot{\epsilon}^p = \dot{\gamma}^p \frac{\sigma}{\sigma^{eq}}\,. \tag{148}$$

The positive scalar $\dot{\gamma}^p$ is not determined by the coherency condition on the criterion (147) but by a differential equation, such as (74), which is rewritten more explicitly as

$$\dot{\gamma}^p = \begin{cases} \dfrac{\sigma^e - \sigma_Y(\gamma^p)}{\eta} & \text{if} \quad \sigma^e \geq \sigma_Y(\gamma^p)\,, \\ 0 & \text{otherwise}\,. \end{cases} \tag{149}$$

The stress point is indeed outside the elastic domain and the distance, in unit of stress, to this domain drives the magnitude of the plastic strain rate. The scalar η in (149) is the viscosity (unit: Pa.s).

Assume known at time t_n, the stress σ_n, the plastic deformation γ_n^p as well as the proposed increment in deformation $\Delta\epsilon$. We seek to determine the stress σ_{n+1} and the plastic strain γ_{n+1}^p at the end of the increment after the time lapse Δt.

1• Transform the equations of 1D visco-plasticity in finite differences following the time discretisation, as discussed after equations (76), for the case of the backward-Euler scheme.

2• What is the elastic predictor for the update of the stress ?

3• What is the condition for the predictor stress to be also the final stress, at the end of the increment ?

4• What is the correction in stress due to the plastic deformation ? In responding to this question, make sure to highlight the structure of the radial return, presented in the main text for strain-rate independent plasticity.

7.3 Problem : Spectral Elements.

The objective of this problem is to construct the 1D version of the spectral elements which are used in the applications presented in section 6. These elements are particularly appreciated for solving wave propagation problems in view of the ease to enrich their kinematics to capture high frequencies. To illustrate the construction of the spectral elements, consider the differential equation

$$u'' + u = x \quad \forall x \in [0; L] \quad \text{with} \quad u(0) = u(L) = 0\,, \tag{150}$$

which is typical of a vibration problem.

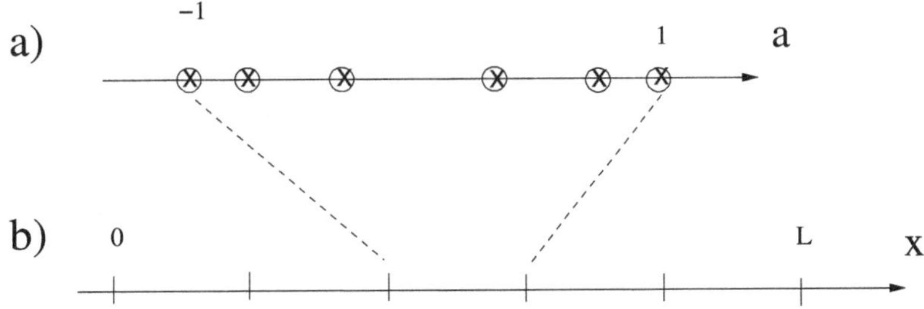

Figure 23. *A* spectral element of order $m = 6$ in its reference configuration a), and the physical domain which is discretised. Note that the nodes (circles) and the Lobatto's quadrature points (crosses) share the same position.

1• What is the weak, symmetric formulation of this problem ? For this purpose, multiply the equation (150) by the test function $v(x)$, with the boundary conditions $v(0) = v(L) = 0$, and integrate over the physical domain. Displacement $u(x)$ and test function $v(x)$ have a spatial interpolation with the same shape functions.

The Lobatto quadrature, discussed in section 4 is selected. The quadrature rule over the interval $[-1; 1]$ is such that the first and the last of the m points are at -1 and + 1, respectively. The positions of these points are presented in Figure 23a for the case of $m = 6$. The second feature of the spectral element is that the number of nodes is equal to the number of quadrature points. Moreover, the nodes and the quadrature points share the same position with a linear mapping between the reference and the physical domain, as illustrated in Figure 23b. The resulting element is not isoparametric. Shape and test functions are then polynomial of order $m-1$, denoted $l_i(a)$ for $a \in [-1;1]$ and $i = 1, ..., m$.

2• What is the value of the shape function $l_i(a)$ at the node a_j to respect the classical conditions on shape functions discussed in section 5 ? Starting from these properties, construct the general form of the function $l_i(a)$ in the reference domain and compute its derivative $l'_i(a)$, for the arbitrary integer m.

3• What are the the components of the local mass matrix ($m \times m$) defined by $M^e_{ij} = \int_{\Omega^e} l_i l_j da$ for an element covering the segment of length L^e ? What is the remarkable property which is observed ?

The domain $[0; L]$ is now discretised with n elements of equal length L^e and the weak formulation obtained in question 1 is used.

4• Give the expressions for the components h^e_{ij} of the local matrix which is the dynamic stiffness, without introducing the numerical quadrature. Give also the the component f^e_a of the local force vector, after quadrature.

The global system of equations results from the contribution of each element. We suppose $n = 3$ and $m = 3$. The global nodal unknowns are noted from U_1 to U_7.

5• What is the global system $[H]$ and of the global force vector $\{F\}$ resulting from the contribution of the three elements.

It is observed in the $n(m-1)-1$ global system, that the unknowns associated to nodes within the element, and not on the element boundary, are only related to the unknowns found on the boundary of the same element. It it thus possible to eliminate these internal unknowns and to deal with each element having m local nodes as an element having only two global nodes.

6• If these internal nodes are eliminated at the local level, what is the size of the global system to be solved in our particular example ($n = 3$ and $m = 3$) ?

8 Solution Hints

8.1 Exercise 1: the Patch Test.

1• The displacement field $u_i = \alpha_i + \beta_{ij} x_j$ is kinematically admissible since continuous. The linearised deformation $1/2(\beta_{ij} + \beta_{ji})$ is homogeneous and leads to a uniform stress field, which is statically admissible. The displacement proposed is thus the exact solution to the problem.

2• The connectivity table is given in Table 6.

Element	local 1	local 2	local 3	local 4
I	1	2	5	4
II	2	3	6	5
III	4	5	8	7
IV	5	6	9	8

Table 6. Connectivity table for the mesh in Figure 22a.

3• If the displacement field leads to a homogeneous deformation, then the numerical interpolation should reproduce this exact solution since complete polynomials were used. This is certainly true for a single element but also for a patch of elements where the exact displacement is imposed at the boundary. The displacement at the 5th node is

$$u_i = \alpha_i + \beta_{ij} x_j^{(5)} .$$

4• The exact deformation is homogeneous. The calculated deformation at any quadrature point should be the same

$$\epsilon_{ij} = \frac{1}{2}(\beta_{ij} + \beta_{ji}) ,$$

within numerical accuracy.

5• The condition presented in question 3 applies also to the mesh based on nine-noded elements. The displacements at the internal nodes are still given by (146). The connectivity table is given in Table 7.

Element	1	2	3	4	5	6	7	8	9
I	1	3	13	11	2	8	12	6	7
II	3	5	15	13	4	10	14	8	9
III	11	13	23	21	12	18	22	16	17
IV	13	15	25	23	14	20	24	18	19

Table 7. Connectivity table for the mesh in Figure 22b.

It is often convenient, during the early phase of the validation of a code development, to check that an irregular mesh, as in Figure 22b, does provide the exact, uniform solution for linear or non-linear problems.

8.2 Exercise 2: Visco-Plasticity of a 1D Von Mises Material

1• The backward-Euler algorithm for the update is

$$
\begin{aligned}
\sigma_{n+1} &= \sigma_n + E(\Delta\epsilon - \Delta\epsilon^p)\,, \\
\Delta\epsilon^p &= \Delta\gamma^p \frac{\sigma_{n+1}}{\sigma^{eq}_{n+1}}\,, \\
\sigma^e_{n+1} &= |\sigma_{n+1}|\,, \\
\Delta\gamma^p \frac{\eta}{\Delta t} &= \sigma^e_{n+1} - \sigma_Y(\gamma^p_{n+1}) \quad \text{if} \quad \sigma^{eq}_{n+1} > \sigma_Y(\gamma^p_{n+1})\,, \\
&= 0 \quad \text{otherwise}\,.
\end{aligned}
\tag{151}
$$

2• The elastic predictor assumes that no plastic straining occurs during the increment:

$$
\begin{aligned}
\sigma_* &= \sigma_n + E\Delta\epsilon\,, \\
\Delta\epsilon^p_* &= 0\,, \\
\sigma^e_* &= |\sigma_*|\,, \\
\Delta\gamma^p_* &= 0\,.
\end{aligned}
\tag{152}
$$

3• We have to make sure that the condition $\sigma^e_* \leq \sigma_Y(\gamma^p_*)$ is respected.

4• If this last condition is not respected, then plastic flow occurs. The stress at the end of the increment is provided by

$$
\begin{aligned}
\sigma_{n+1} &= \sigma_* - E\Delta\epsilon^p\,, \\
\Delta\epsilon^p &= \Delta\gamma^p \frac{\sigma_{n+1}}{\sigma^e_{n+1}}\,, \\
\sigma^e_{n+1} &= |\sigma_{n+1}|\,, \\
\Delta\gamma^p \frac{\eta}{\Delta t} &= \sigma^e_{n+1} - \sigma_Y(\gamma^p_{n+1})\,.
\end{aligned}
\tag{153}
$$

Combining (153a) and (153b) one obtains

$$
\sigma_{n+1}\Big(1 + \Delta\gamma^p \frac{E}{\sigma^{eq}_{n+1}}\Big) = \sigma_*\,,
\tag{154}
$$

and also

$$
\sigma^{eq}_{n+1} + \Delta\gamma^p E = \sigma^e_*\,.
\tag{155}
$$

Equation (153d) then becomes

$$\Delta\gamma^p(\frac{\eta}{\Delta t} + E) + \sigma_Y(\gamma_n^p + \Delta\gamma^p) = \sigma_*^e\,. \tag{156}$$

This non-linear equation is solved by Newton-Raphson for the positive scalar $\Delta\gamma^p$. The final, equivalent stress is then $\sigma_{n+1}^{eq} = \sigma_*^{eq} - \Delta\gamma^p E$, according to (155), and the final update is done according to (154)

$$\sigma_{n+1} = \frac{1}{1 + \frac{E\Delta\gamma^p}{\sigma_{n+1}^e}}\sigma_*\,. \tag{157}$$

We have indeed found the structure corresponding to the radial return defined in the main text for rate-independent plasticity.

8.3 Problem: the Spectral Elements.

1• Multiply (150) by the test function $v(x)$, which satisfies $v(0) = v(L) = 0$, and integrate the result over the domain of study

$$\int_0^L (u'' + u - x)v\,dx = 0\,.$$

Integrate the first term by parts and take into account the boundary conditions on the test function to obtain the symmetric weak formulation

$$\int_0^L -u'v' + uv - xv\,dx = 0\,.$$

2• The shape function $l_i(a)$ is zero at every node j different from i and takes the value one at node i. Theses conditions are equivalently written as $l_i(a_j) = \delta_{ij}$. For m quadrature points, the shape function $l_i(a)$ must be zero at $m-1$ nodes. It is thus proportional to the product

$$\prod_{\alpha=1,\neq i}^{m} (a - a_\alpha)\,.$$

It suffices now to normalize this product such that the function is equal to one at node i

$$l_i(a) = \frac{\prod_{\alpha=1,\neq i}^{m} (a - a_\alpha)}{\prod_{\alpha=1,\neq i}^{m} (a_i - a_\alpha)}\,.$$

The first derivative of this function is

$$l_i'(a) = \sum_{\beta=1}^{m} \frac{\prod\limits_{\alpha=1,\neq i,\neq\beta}^{m} (a - a_\alpha)}{\prod\limits_{\alpha=1,\neq i}^{m} (a_i - a_\alpha)} .$$

3• The spatial interpolation is linear and the Jacobian of the transformation from the reference to the physical domain is the constant $L^e/2$. The computation of the mass matrix is done in the computational domain from

$$\frac{L^e}{2} \int_{-1}^{+1} l_i(a) l_j(a) da .$$

It is approximated by the following weighted sum according to the Lobatto's quadrature

$$\frac{L^e}{2} \sum_{\alpha=1}^{m} l_i(a_\alpha) l_j(a_\alpha) w_\alpha .$$

The shape functions are evaluated at the nodes which are also the quadrature points so that this sum simplifies to

$$\frac{L^e}{2} \sum_{\alpha=1}^{m} \delta_{i\alpha} \delta_{j\alpha} w_\alpha$$

or furthermore

$$\frac{L^e}{2} \delta_{ij} w_j ,$$

without summing over j despite the repetition of this index. The mass matrix is thus diagonal. This remarkable property applies also in 2D and 3D. The computation of the accelerations is then very efficient computerwise (see, for example, Hughes (1987), for further discussion).

4• The local, dynamic stiffness matrix has the following components

$$\frac{L}{2} \delta_{ij} w_j - \frac{2n}{L} \int_{-1}^{+1} l_i'(a) l_j'(a) da ,$$

without any sum despite the repeated j. The second term of this expression is approximated by Lobatto's quadrature. This matrix is not diagonal. The local force vector has for components

$$\frac{L}{2n} \int_{-1}^{+1} l_i(a) x da .$$

The quadrature provides

$$\frac{L}{2n} w_i x_i \,,$$

with no sum over i.

5• The global system of equations accounting for the boundary conditions is $[H]\{U\} = \{F\}$ or more explicitly

$$\begin{bmatrix} h_{22}^{(1)} & h_{23}^{(1)} & 0 & 0 & 0 \\ h_{32}^{(1)} & h_{33}^{(1)} + h_{11}^{(2)} & h_{12}^{(2)} & h_{13}^{(2)} & 0 \\ 0 & h_{21}^{(2)} & h_{22}^{(2)} & h_{23}^{(2)} & 0 \\ 0 & h_{31}^{(2)} & h_{32}^{(2)} & h_{33}^{(2)} + h_{11}^{(3)} & h_{12}^{(3)} \\ 0 & 0 & 0 & h_{21}^{(3)} & h_{22}^{(3)} \end{bmatrix} \begin{Bmatrix} U_2 \\ U_3 \\ U_4 \\ U_5 \\ U_6 \end{Bmatrix} = \tag{158}$$

$$\frac{L}{6} \begin{Bmatrix} w_2^{(1)} x_2 \\ (w_3^{(1)} + w_1^{(2)}) x_3 \\ w_2^{(2)} x_4 \\ (w_3^{(2)} + w_1^{(3)}) x_5 \\ w_2^{(3)} x_6 \end{Bmatrix} .$$

6• The second nodal unknown of each element can be eliminated. The global system of five equations is thus reduced to a system of two equations only. This procedure is often referred to as the static condensation.

The generalization of the construction just presented to 2D and 3D is straightforward. The shape functions for the multi-dimensional setting are simply the product of the 1D functions derived here. The interested reader is referred to Komatitsch and Vilotte (1998) for further discussion in the dynamic context. It is this class of elements which are used at the end of section 6 for quasi-static applications.

Acknowledgments: This chapter has benefited greatly from the detailed reading of G. Kampfer (ENS) and from the comments of J.R. Raphanel (Ecole Polytechnique).

Bibliography

Babuška, I. and T. Strouboulis (2001). *The finite element method and its reliability.* Oxford University Press, Oxford.

Bazeley, G.P., Cheung, Y.K., Irons, B.M. and O.C. Zienkiewicz (1965). Triangular elements in plate bending – conforming and nonconforming solutions. In: *Proceedings of the first conference on Matrix Methods in Structural Mechanics*, Wright-Patterson AFB, Ohio.

Becker, E. (1981). *Finite Elements.* Prentice-Hall.

Belytschko, T., Kam Liu, W. and B. Moran (2000). *Nonlinear finite elements for continua and structures.* John Wiley & Sons, Chichester, England.

Bonnet, M. and A. Frangi (2005). *Analyse des solides déformables par la méthode des éléments finis.* Edition Ecole Polytechnique, Palaiseau, France.

Chadwick, P. (1976). *Continuum mechanics - Concise theory and problems.* G. Allen & Unwin, Dover Edition 1999.

Ciarlet, P.G. (1978). *The finite element method for elliptic problems.* North-Holland, New York.

GiD (2009). The personal pre and post processor, International Center for Numerical Methods in Engineering, Barcelona (www.gidhome.com).

Hildebrand, F.B. (1974). *Introduction to numerical analysis.* Second edition, Mc Graw-Hill Inc., Dover Edition 1987.

Hill, R. (1950). *The mathematical theory of plasticity.* Oxford University Press, Oxford.

Hughes, T.J.R. (1987). *The finite element method, Linear static and dynamic finite element analysis.* Prentice-Hall Inc., Englewood Cliffs, New Jersey.

Hughes, T.J.R. and J.C. Simo (1998). *Computational Inelasticity.* Series - Interdisciplinary Applied Mathematics, Springer-Verlag, New York.

Johnson, C. (1987). *Numerical solution of partial differential equations by the finite element method.* Cambridge University Press, Cambridge.

Komatitsch, D. and J.-P. Vilotte (1998). The spectral element method: an efficient tool to simulate the seismic response of 2D and 3D geological structures, *Bull. Seism. Soc. Am.*, 88, 368–392.

Malvern, L.E. (1969). *Introduction to the mechanics of a continuous medium.* Prentice-Hall.

Massin P., Triantafyllidis, N. and Y.M. Leroy (1996). Stability of a density-stratified two-layer system. *C.R. Acad. Sci. Paris, Tectonique,* Série IIa, 322, 407–413.

Ortiz, M., Leroy, Y. and A. Needleman (1987). A finite element method for localized failure analysis. *Comput. Methods in Appl. Mech. Engrg.*, 61, 189-214.

Reddy, B.D. (1991). *Introductory functional analysis.* Texts in Applied Mathematics 27, Springer.

Salençon, J. (2001). *Handbook of continuum mechanics, General concepts and thermoelasticity.* Springer.

Simo, J.C. and R.L. Taylor (1985). Consistent tangent operators for rate independent elastoplasticity. *Comput. Methods in Appl. Mech. Engrg.*, 51, 177–208.

Strang, G. (1972). Variational crimes in the finite element method. In: *The mathematical foundations of the finite element method with applications to partial differential equations*, edited by A.K. Aziz, Academic Press, New York, 689–710.

Strang ,G. and G.J. Fix (1973). *An analysis of the finite element method.* Prentice-Hall Inc., Englewood Cliffs, New Jersey.

Timoshenko, S.P. and J.N. Goodier (1934). *Theory of Elasticity.* McGraw-Hill.

Tvergaard, V., Needleman, A. and K.K. Lo (1981). Flow localization in the plane tensile test. *J. Mech. phys. Solids*, 29, 115–142.

Zienkiewicz, O.C. (1977).*The finite element method.* Third, expanded and revised edition of *The finite element method in engineering sciences.* McGraw-Hill.

Zeitfracht Medien GmbH
Ferdinand-Jühlke-Straße 7
99095 Erfurt, Deutschland
produktsicherheit@kolibri360.de